The Quantum World Unveiled by Electron Waves

The Quantum World Unveiled by Electron Waves

Akira Tonomura

World Scientific
Singapore • New Jersey • London • Hong Kong

Published by

World Scientific Publishing Co. Pte. Ltd.

P O Box 128, Farrer Road, Singapore 912805

USA office: Suite 1B, 1060 Main Street, River Edge, NJ 07661

UK office: 57 Shelton Street, Covent Garden, London WC2H 9HE

British Library Cataloguing-in-Publication Data
A catalogue record for this book is available from the British Library.

THE QUANTUM WORLD UNVEILED BY ELECTRON WAVES

ISBN 981-02-2510-5 ✓

Printed in Singapore by Eurasia Press Pte Ltd

Preface

This book is an interesting personal account in non-technical terms of the research experience in basic sciences by one of the world's greatest experts on electron waves. It covers many facets of his beautiful works in the last thirty years.

Electrons, nuclei and photons are the particles that are the main subjects of study of twentieth century physics. Based on such studies there was remarkable success in the manipulation of the motions and intereactions of these particles, giving rise in turn to the new technologies of the twentieth century. This success is obviously going to be continued into the next century, with profound implications for the economies of the industrialized world.

Matter is made of electrons and nuclei. Since nuclei are heavy and tightly bound, and cannot be easily broken apart, properties of matter are usually determined by the motions of electrons. This explains, *a priori*, why the study of the properties of the motions of electrons is so important. Dr. Tonomura's success began with his ability to produce coherent motion of electrons. This success led him to previously unattainable precision in many beautiful experiments.

The personal touch adds flavor to the book. Although the author often has to venture into technical details in describing some developments, it is worthwhile for the reader to wade through such passages of heavy reading since the author does succeed in describing the great intellectual and practical excitement involved in such developments, bringing to life the feelings of a scientist in the frontiers of research.

Chen Ning Yang

Prologue

I wrote this monograph to excite people's curiosity about the strange behavior of electrons seen in the microscopic world, behavior quite different from those of common objects we see in our daily life. The story begins with my Friday Evening Discourse at the Royal Institution in Great Britain on November 4, 1994. This event gave me my first chance to talk to laymen while using demonstration experiments, and it made me think about "science in the society" and finally made me take up my pen. I would not like to simply describe a dry collection of knowledge but instead to tell you of the enthusiasm, joy, and difficulty of my own research. Among the fields of science treated in this book are quantum mechanics, superconductivity, electron microscopy, holography, magnetism, and unified theories. These are the fields I explored using electron waves.

"Why did I write such a book?" Because I want young people to enter the fields of physics where there are still many interesting problems to be solved. We need young people who think logically and deeply. No, such abilities may be cultivated later on. However, I believe those who are most suitable are moved by beauty and think that it is worthwhile even to spend their whole lives to see the beauty of Nature.

In any case, things become difficult when you want to open a new way. Although you might feel uneasy about whether you can do such a thing alone, you might instead become courageous and awaken your adventurous spirit.

When you throw yourself into your curiosity and dare to dream, there will often be times you are not successful. Even in that case, you should never get discouraged but forget the failures soon, thinking that this is due to your shallow considerations. But once in a while when you succeed and peep into the beauty of Nature, you will be indescribably delighted. I wrote this book expecting lots of readers to think, "If there is such a way, I want to step forward on that way!".

Contents

Contents

Chapter 1

MAGNETIC LINES OF FORCE

My Friday Evening Discourse started with a demonstration of magnetic lines of force.

Magnetic Lines of Force

Let us place a horseshoe magnet on my desk, put a glass plate over the magnet, and then sprinkle iron filings on the plate. Magnetic lines of force begin to appear on the glass plate (Fig. 1). Fine iron particles are connected in lines to form a pattern as if something were flowing from one pole of the magnet to the other. These are

Fig. 1. Iron filings displaying magnetic lines of force leaking from a horseshoe magnet.

the *magnetic lines of force*. The concept of lines of force was devised in 1831 by Michael Faraday at the Royal Institution in Great Britain (Fig. 2).

Faraday recognized in 1851 that lines of force flow not only outside the magnet but also *inside* it. If you cut the magnet into two pieces, you will find that magnetic

1

Fig. 2. Michael Faraday (1791–1867) (by courtesy of the Royal Institution, Great Britain).

lines of force come out from the cross-section. To describe the magnetic state inside and outside a magnet, Faraday proposed the concept of *magnetic induction*. A stream of the magnetic induction is like a flux having neither sources nor sinks, and the strength of the magnetic induction reflects the density of magnetic lines of force.

The magnetic lines of force, or more precisely, the lines of magnetic induction, shown in Fig. 1 provide a picture of the magnetic induction proposed by Faraday. But in this picture we can only see the lines outside the magnet. "How can we see them flowing inside the magnet?" you might ask. Of course, no one can sprinkle iron filings inside the magnet. But, as I will show you later, the magnetic lines of force inside the magnet can be brought out by sprinkling *electrons* instead of iron filings.

By using electrons, we can observe magnetic lines of force in the microscopic world quantitatively. Being able to see magnetic lines of force on a microscopic scale greatly helps us understand, and in fact control, what happens in such useful materials as magnetic materials and superconductors.

Magnetic Lines Drawn with Electrons

Let me first show you an example of magnetic lines of force in the microscopic region. We look at a *magnetic bubble memory device*, which is used in a telephone

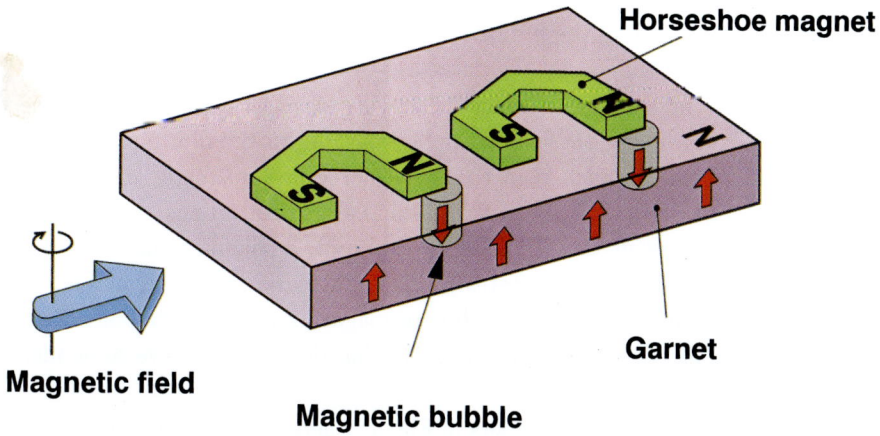

Fig. 3. Magnetic bubble memory.
   The magnetic lines of force leaking from tiny horseshoe magnets make magnetic bubbles move by the rotation of the applied magnetic field.

switchboard. Ladies know about garnet jewels, and there is a magnet material whose crystal structure is the same as that of garnets. When a thin film is made of it, the N pole appears on one surface of the film and the S pole on the other (Fig. 3). When a magnetic field is applied in the opposite direction to the magnetic field inside the film, tiny opposite magnets are produced in the film. Since they look like bubbles as you see in Fig. 3, they are called *bubble domains.* In this memory device, information is recorded in such a way that the presence of the bubbles means "yes," and the absence of the bubbles means "no." When we want to read

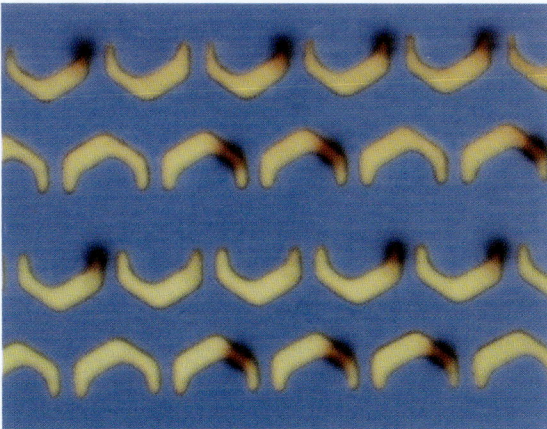

Fig. 4. Tiny horseshoe magnets used for propagating magnetic bubbles.
   The bubbles are visible as slightly dark shadow spots in this frame of the video tape taken by optical polarization microscopy.

the information we push all the bubbles to the right end where the existence or the absence of bubbles is decided by the detector.

Thousands of tiny horseshoe magnets are arranged in a plane just above the garnet film. When a magnetic field is applied parallel to the film plane, the horseshoe magnets (which, distinct from the usual horseshoe magnets, are not permanent magnets, and are made of a material similar to iron filings) are magnetized to have N and S poles in the direction of the applied magnetic field. When the direction of the magnetic field rotates in the film plane as shown in Fig. 3, the positions of the N and S poles move, thus propagating magnetic bubbles from left to right towards the detector. In this way, this device can serve as a memory device.

You can see tiny horseshoe magnets in Fig. 4, where magnetic bubbles appear as dark circular shadows attached to poles. Magnetic bubbles can be seen by using an optical polarization microscope. It is the magnetic lines of force from the horseshoe magnets that make these bubbles move.

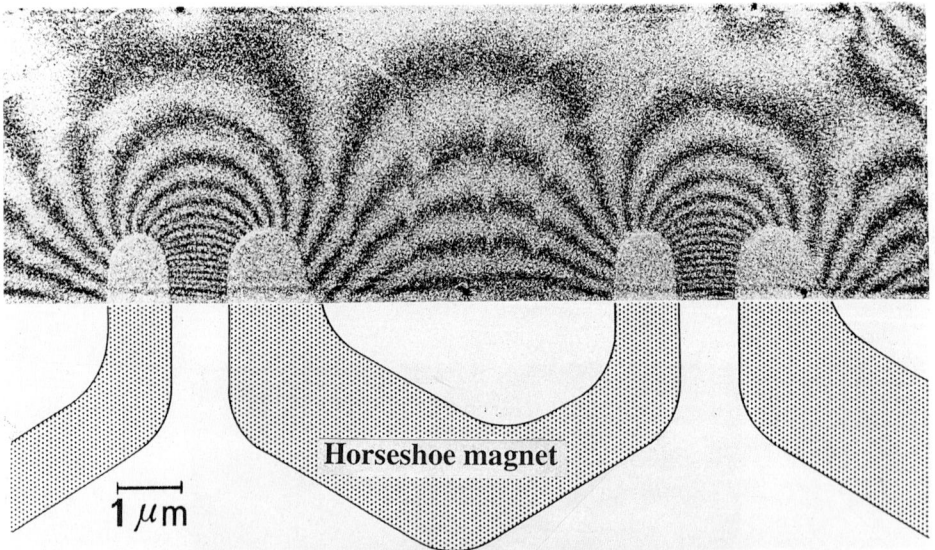

Fig. 5. Magnetic lines of force from horseshoe magnets observed by using electron waves.

The picture in Fig. 5 shows the magnetic lines of force leaking from the tiny horseshoe magnets. It is to be noted that these magnets are extremely small, the width of the poles being 1 μm, or only one hundredth the diameter of a human hair. "Was this picture made by sprinkling iron filings?" No!! These magnets are so small — even smaller than a single grain of the iron filings that were used to form the magnetic lines of force in Fig. 1 — that their magnetic lines cannot be visualized by sprinkling iron filings.

This picture was instead made by using *electron waves*. We sprinkled *electron waves* instead of iron filings over the magnets. And then, we used what is called the *interference* of electron waves.

"What are the *electron waves*?" "What is the *interference*?" Electron interference is the most important tool that I use in my research and is also the most important concept in this book. I would therefore like to start by demonstrating what waves are.

Chapter 2

WHAT ARE WAVES?

Waves on a String

Let me first demonstrate interference by using waves on a string as the simplest example of a one-dimensional wave. The apparatus shown in Fig. 6 was developed for demonstrating waves at my Friday Evening Discourse at the Royal Institution.

Fig. 6. Demonstration of string waves.
What happens to string C when two strings A and B are made to oscillate horizontally?

(The details of the apparatus are illustrated in Fig. 7 for your reference.) Two strings of the same length are tied to a single string, and in Fig. 6 you can see the letter Y with a blue branch (A) on the top extending toward your right, a white branch (B) on the bottom also extending toward your right, and the red leg (C) extending toward your left. I use this set-up to explain what happens when two waves meet.

First, let us vibrate the end of the blue branch. The wave generated at the end travels from the blue branch to the red leg [Fig. 8(a)]. The string vibrates most strongly at the green ball 2 on the red string, and there is no vibration at the yellow ball 1. The point of no vibration is called a "node." In this case, the node happens

6

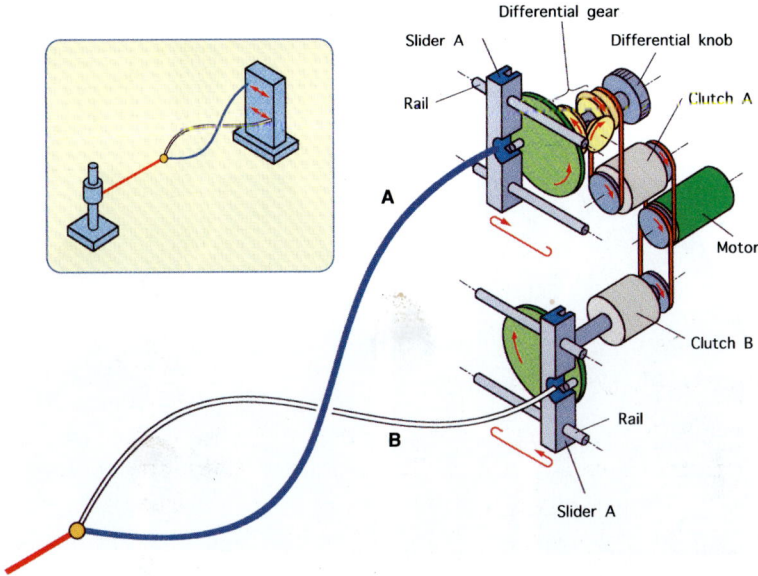

Fig. 7. Apparatus for demonstrating the interference of two string-waves.
Two strings can be made to oscillate independently; sometimes *in phase*, sometimes *out of phase*.

to be located at the knot of the strings, the center of the Y. Under this condition, the wave travels to the red leg of the Y, while the white branch remains still.

Now let us generate a wave at the end of the white branch. Again the wave propagates to the red string [Fig. 8(b)], causing it to vibrate exactly as in the previous conditions. This time the blue branch remains still.

"What happens when we generate two waves on both the blue and the white strings at the same time?" Let us first vibrate the ends of both branches *synchronously*. The two waves coming from the two branches overlap on the leg of the Y and the crests of the two waves meet at the green ball 2, causing the vibration to become stronger [Fig. 8(c)]. The amplitude of the vibration at this green ball has doubled. This is the result of a wave interfering with another wave whose phase is the same. In this case, the two waves are said to be *in phase*.

"What happens if the two waves are *out of phase*?" Let's vibrate the ends of the two branches in opposite directions. You see in Fig. 8(d) that the two branches, blue and white, vibrate in opposite directions (*out of phase*). Then, the vibration on the leg of the Y disappears because the crest of the wave coming from one branch meets the trough of the wave coming from the other branch. The opposite vibrations cancel each other.

When two waves meet, they interfere. You have just seen the interference between the blue wave and the white wave propagating into the red string. When the two waves are *in phase* the vibration is doubled, and when the two waves are made to meet *out of phase* they cancel each other.

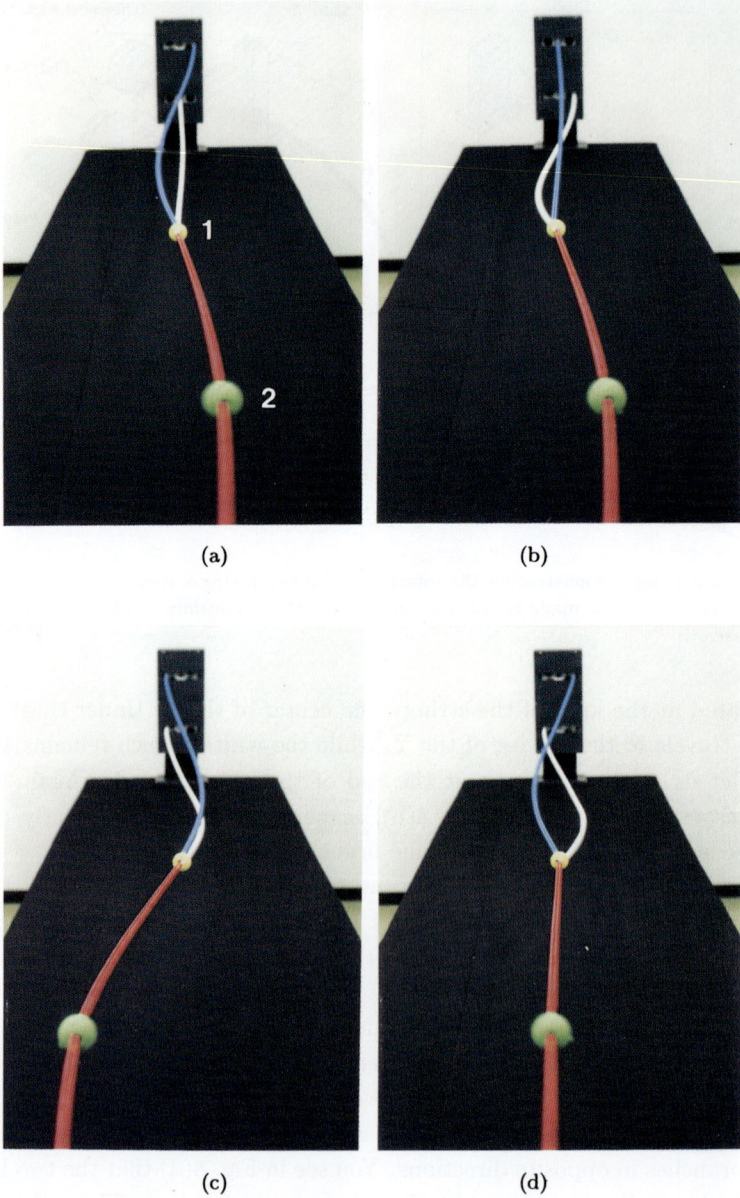

Fig. 8. Overlap of two waves on strings.
    (a) What happens when the blue string A is made to vibrate?
    (b) What happens when the white string B is made to vibrate?
    (c) What happens when the two strings are made to vibrate *in phase*?
    (d) What happens when the two strings are made to vibrate *out of phase*?

You can tell the relative phase of one wave with respect to the other just by watching the red string, or *the result of interference*. If you know that you have a strong vibration on the red string, you can tell that there is a white wave *in phase* with the blue wave. And if you don't see vibrations on the red string, you can tell that the blue wave and the white wave are completely *out of phase*.

This is an example of interference for waves in one dimension. Now I'll show you the interference of waves in two dimensions by using a water tank.

Water Waves

Here is a transparent plastic tank full of water (Fig. 9). Under the tank we put an overhead projector so that we can see how waves propagate on the surface of the water by looking at their projection on the screen.

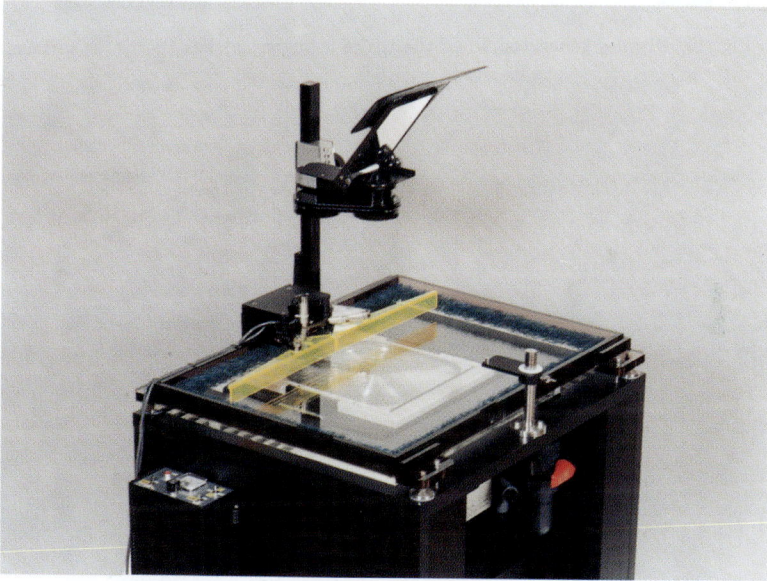

Fig. 9. Ripple tank.

When you tap the water surface with a finger, you generate a wave and you can see the ripples, or a two-dimensional wave, propagate (Fig. 10). You can also generate waves by using the bar at the end of the tank, which vibrates up and down, hitting the water surface (Fig. 11). The crests and troughs propagate one following the other. The contours of the wave crests are called *wavefronts*. The wave proceeds perpendicular to the wavefronts (Fig. 12). In this case, the wavefronts are straight lines and such a wave is called a *plane wave*.

The crests of the waves shine brightly, forming the bright horizontal lines on the screen. In between the crests there are troughs forming dark lines.

Fig. 10. Ripples generated when the tip of a finger touches the water surface.

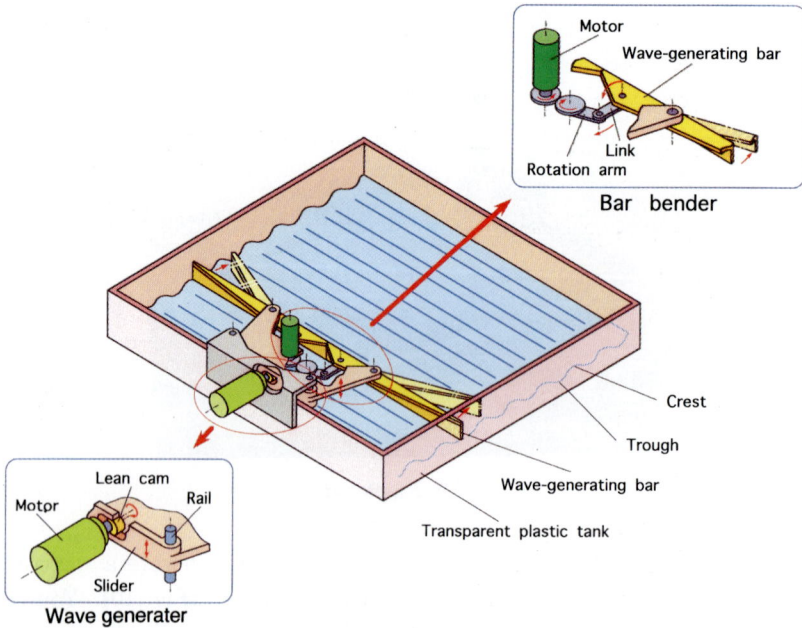

Fig. 11. Transparent water tank for observing ripples on the water surface.
When the motor on the left is turned on, the lean cam makes the slider oscillate up and down. As the wave-generating bar also oscillates up and down to tap the water surface, plane waves are generated. When the other motor above is turned on, the wave-generating bar bends at its center and becomes V-shaped. If the bar oscillates up and down, two plane waves are generated and travel in two different directions. An interference pattern can be observed where the two plane waves overlap.

Fig. 12. A plane wave produced when the bar oscillates.

Interference of Two Waves

Now let us perform an interference experiment using the two-dimensional waves on the water surface. For this experiment we bend the wave-generating bar into a V-shape. This can be done by starting the bending motor (see Fig. 11). Two plane waves are then generated and propagate in two directions. Since waves propagate perpendicularly to their wavefronts, the two waves overlap in the middle of the tank and interfere with each other. You can see in Fig. 13(a) a new pattern in the region where the two waves overlap. This is the result of *interference*.

This pattern looks complicated, but can be interpreted simply. The bright region indicated by the red spot in Fig. 13(a) is the place where the crest of one wave meets the crest of the other to form a higher crest. The dark region just below is the place where the trough meets the trough to form a deeper trough. This again illustrates the interference of waves, but in two dimensions. The water surface oscillates up and down vigorously at these places along the central vertical line in the figure, since there the two waves overlap crests to crests or troughs to troughs, that is to say, *in phase*.

Recall the string experiment shown in Fig. 8(c): when the two waves overlap in phase, the red string oscillates vigorously. The central vertical line in Fig. 13(a) is the same distance from the two bars, and the two waves overlap *in phase* there. The water surface oscillates up and down vigorously along this line. You can find several more such vertical lines along which the vibration is enhanced. Between these lines, on the other hand, the crests meet the troughs, and vice versa, to annihilate each other, and there is no vibration there.

Now let's consider how the oscillation changes as we go in the horizontal direction in Fig. 13(a). Strong vibrations alternate periodically with no vibrations.

"What happens if you place an object in one of the two waves before they meet?" As can be seen in Fig. 14, the object distorts the wavefronts of one wave.

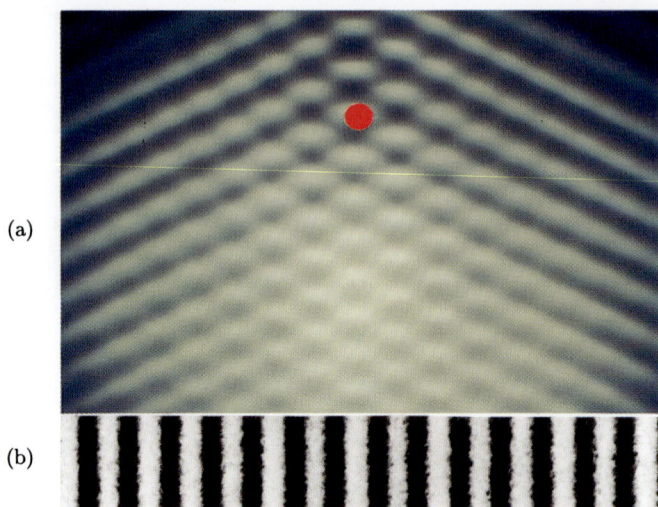

(a)

(b)

Fig. 13. An interference pattern between two plane waves.
(a) Wavefronts on a water surface
(b) Interference fringes of light
The bright spot indicated by the red disc is located where the crest of one wave meets the crest of the other wave to form a higher crest. Similarly the dark spot just below it indicates a deeper trough. The water surface oscillates vigorously along the central vertical line in picture (a), since the two waves overlap in phase. Such lines of enhanced oscillation appear periodically.

Fig. 14. Wavefronts distorted by an object.
Wavefronts of one plane wave are disturbed by the object which is placed in the water. The interference fringes are also modulated.

Light Waves

If you want to observe the interference of light waves, you need to use a photographic film because you cannot see the light wavefronts the way you can see water wavefronts. We get strong exposure at the places where the vibrations are strong and no exposure between those places. Hence, we get a regular array of black and white stripes like the picture in Fig. 13(b). These stripes are called *interference fringes*, and they are formed when two plane waves overlap.

When you place an object in one wave (Fig. 14), the wavefront distortion will show up as irregularities in the interference fringes. If you could record the intensity of the distorted interference patterns, you could get information about the distortion of the wavefronts. You can easily understand why. In the interference pattern, the phase shift of the distorted wave relative to the reference wave is recorded. You saw in the string experiment that we could know the phase shift of one wave relative to the other from the oscillation of the red string, the result of the interference. The same thing happens with light waves.

It was Dennis Gabor who noticed this in 1948 and devised *holography*, the method for reconstructing wavefronts. Gabor attempted to record on film, as a *hologram*, the wavefronts of electrons scattered by an object, and to reconstruct an optical analog of the electron wavefronts. He wanted to break through the resolution limit of electron microscopes, compensating for the inevitable aberrations in the electron lens by using optical techniques on an optical bench.

You may say that electrons are not waves but particles, with which interference fringes cannot be formed. However, this is not the case. Electrons *sometimes* behave as if they were waves, and as such they play the starring role in the field I am going to talk about from now on. Therefore, I would like to describe the story of electrons in the next chapter.

Chapter 3

INTERFERING ELECTRONS

We have talked about the properties of the waves using string waves and water waves. It is time to talk about the electrons that play the main role throughout this book.

Discovery of Electrons

Electrons were discovered in 1897 by J. J. Thomson in Great Britain. He settled the long dispute concerning what *the cathode ray* was. He found experimental evidence that the cathode ray is a stream of negatively charged tiny particles. Up to that time, some people had asserted that the cathode ray emitted from the cathode in a vacuum tube must be some kind of waves, like electromagnetic waves, and others had asserted that they must be particles.

Thomson also predicted that atoms were composed of these electrons since electrons had the same ratio of electric charge to mass whatever materials they emerged from. Since an atom is neutral, there must be a positive charge inside the atom. Thomson's atomic model was a positively charged pudding with electrons embedded like raisins. His cathode ray was a stream of raisins coming out from the pudding.

The Structure of an Atom

Thomson revealed that electrons were particles. However, his pudding model turned out to be inconsistent with the new experimental result on the scattering of heavy α-particles incident onto gold films. This was because α-particles were back-scattered with a probability much larger than that expected from a pudding model, through which all the α-particles should pass without deflection.

In 1911, E. Rutherford showed that the number of back-scattered α-particles could not be accounted for by using the pudding model and that the positive charge must be distributed not uniformly in an atom but concentrated in a small nucleus. Previously in 1904, a nuclear model had been proposed by H. Nagaoka in Japan. But, his nucleus was nearly as large as the atom, so that his model was called Saturnian.

A new atomic model was thus established: electrons revolve around a heavy and small atomic nucleus. It was soon revealed, however, that this model was fatally flawed: revolving electrons radiate light and lose energy. Electrons would thus fall into the nucleus soon, and the atom collapses. This is an inevitable conclusion from the Newtonian mechanics and Maxwell's electromagnetic theory. Therefore atoms could not exist if the Newtonian mechanics and Maxwell equations were assumed to be valid.

Another problem was pointed out: if electrons lose energy, they trace spiral trajectories. Consequently the period of one revolution becomes shorter and shorter. According to the Maxwell theory, the frequency of emitted light is the same as the frequency of revolution, thus making the wavelength of the emitted light shorter. The wavelength of light emitted from atoms must be distributed continuously. However, the fact is different: the observed wavelengths of light from atoms always have definite discrete values.

In 1913 N. Bohr in Denmark solved the riddle of the definite wavelengths of light emitted from atoms by postulating that electrons in an atom move in nonradiating orbits which have discrete energies. When atoms are excited, electrons in a lower-energy orbit are excited to a vacant higher-energy orbit and in due time return to the original orbit emitting light. The energy of the light is given by the difference between the energy levels of the two orbits. Bohr could explain the observed wavelengths of light emitted from atoms by his theory.

Wave Nature of Electrons

"Why are only certain orbits allowed?" A young French prince, L. de Broglie, worked out an answer to this question in 1924 in his dissertation: he suggested that electrons might have wave properties. If electrons formed *standing waves* surrounding the nucleus just like the vibration of strings of a violin, special orbits would be selected automatically. As shown in Fig. 15, a standing wave is a coherent wave around the nucleus. This wave can be formed only when the length of one orbital revolution is equal to one wavelength, two wavelengths, ... and so forth.

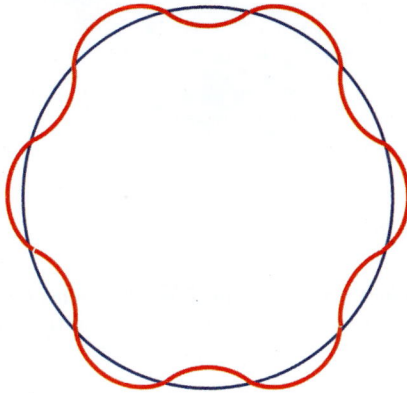

Fig. 15. The electron wave surrounding an atomic nucleus.

Thus, electrons in an atom are allowed to move only in certain orbits, which have discrete energy values.

There was a storm of activity for a few years after that: E. Schrödinger in Austria, W. K. Heisenberg in Germany, and P. A. M. Dirac in Great Britain competed in establishing their own theories of electrons, and the riddles of those days were solved one after the other. The theory thus completed is called *quantum mechanics*, and in the microscopic world inside atoms it replaces the Newtonian mechanics.

Direct Evidence for Electron Waves

"Does the wave nature of electrons appear only in the world of atoms and molecules?" No, physical laws should be consistent in both microscopic and macroscopic worlds. Therefore, free electrons traveling in the vacuum, too, should have wave properties, although their wavelengths are extremely short as compared with the distance they travel. According to the de Broglie relation, the wavelength λ of electrons is determined by their velocity v:

$$\lambda = \frac{h}{mv}, \tag{1}$$

where h and m are Planck's constant and electron mass, respectively 6.6×10^{-34} Js and 9.1×10^{-31} kg. When an electron beam is accelerated by a potential V (volts), then $eV = \frac{1}{2}mv^2$, where e is the charge of an electron and is 1.6×10^{-19} C. Consequently,

$$\lambda = \frac{h}{\sqrt{2meV}} = \sqrt{\frac{150}{V\,(\text{volts})}} \;(\text{Å}). \tag{2}$$

When $V = 100$ kV, λ is 0.04 Å $= 4 \times 10^{-10}$ cm. Since the electron wavelength is so short, interference should occur only in an extremely small region.

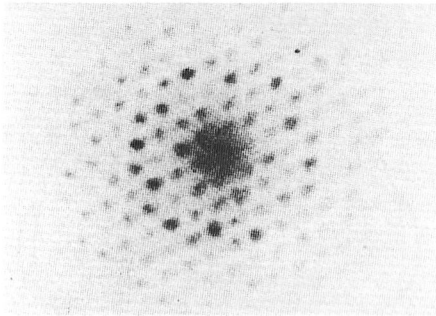

Fig. 16. An electron interference pattern due to crystalline atoms (S. Kikuchi).
   An electron wave is scattered by crystalline atoms and the scattered electron waves interfere with each other to form this diffraction pattern.

In 1927 C. J. Davisson and L. H. Germer in the United States and G. P. Thomson in Great Britain obtained direct evidence for the wave nature of free electrons traveling in a vacuum: Davisson and Germer observed the oscillatory intensity distribution of electrons reflected from surfaces of single crystalline nickel against the angle of reflection, which could be interpreted as the result of the interference of electron waves reflected from regularly arranged atoms. Thomson also observed such an oscillatory intensity of electrons transmitted through a thin polycrystalline film of celluloid as a diffraction ring pattern just like the so-called Debye-Scherrer ring photographed using X-rays. Half a year later, S. Kikuchi in Japan also obtained beautiful electron diffraction patterns from a thin film of a single mica crystal just like Laue spots observed with X-rays, as shown in Fig. 16.

Conditions for Forming Interference Patterns

Interference fringes cannot always be observed with any electron beam. "Under what conditions can interference fringes be observed?" Let us first consider the case when a plane wave is incident onto a screen with two slits [Fig. 17(a)]. When a plane wave is incident *perpendicularly* onto the screen with two slits, cylindrical waves are produced from the two slits to form an interference pattern: Just as in the case of water waves [Fig. 13(a)], the two cylindrical waves overlap *in phase* to oscillate synchronously at the central point O in the observation plane. A nonzero phase difference between the two cylindrical waves is produced at a point which is different from point O, and its value increases with the distance from point O. Every time the phase difference increases by 2π, the amplitude of the oscillation takes the maximum value thus forming interference fringes.

Up to now we have assumed an electron *plane* wave as an incident wave, which corresponds to an electron traveling in a free space with a definite speed and a definite direction. The phase of a plane wave is a constant everywhere in a plane perpendicular to the propagation direction of the wave, and the wave repeats the same pattern at every wavelength in the propagation direction. Therefore, the wave extends to infinity in the propagation direction as well as in a plane perpendicular to it. You may not understand why a single electron is not localized and is anywhere in space. You are right. Such an electron, or an ideal plane wave, however, cannot actually be obtained.

An actual electron is represented not by a plane wave but by a spatially localized *wave packet*. A wave packet consists of plane waves whose wavelengths and propagation directions are nearly equal with small differences. Accordingly, the electron corresponding to a wave packet has spreads, or uncertainties, in its speed and in its direction of motion. Now, let us consider what will happen to the two-slit interference pattern [Fig. 17(a)] when an incident electron has slight uncertainty in its direction of motion. The speed of the electron is assumed to be definite. This wave can be considered to consist of plane waves having different propagation directions. "How is the interference pattern affected when an incident plane wave is tilted by an angle α?" The interference pattern is displaced in proportion to α as illustrated in Fig. 17(b).

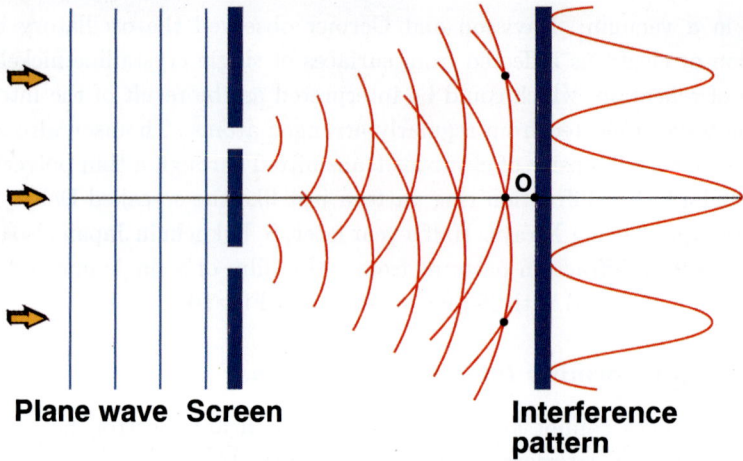

(a)

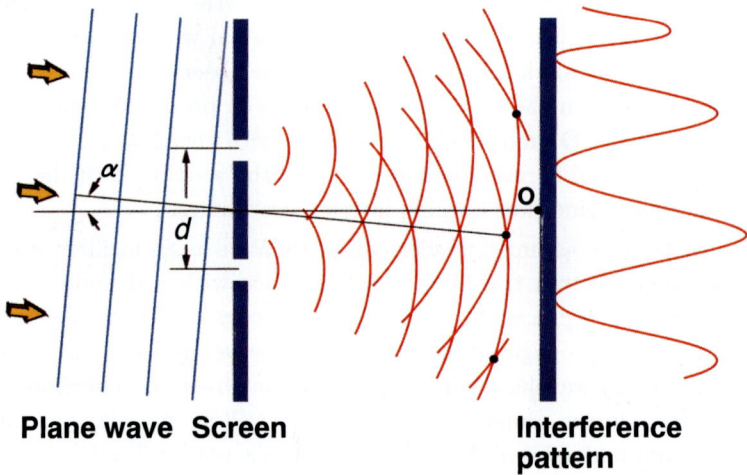

(b)

Fig. 17. Two-slit interference patterns.
  (a) Tilting angle (α) = 0
  (b) Tilting angle (α) $\neq$ 0
  When an incident plane wave is tilted by angle α, the interference pattern is displaced in proportion to α.

This can also be understood by considering the intensity of the interference pattern at the central point O in the observation plane. In Fig. 17(a), two cylindrical waves are *in phase* at O, and consequently the intensity has the maximum value there. In Fig. 17(b), a relative phase shift of θ is produced between the two waves at point O. This shift θ is equal to the phase difference of an incident tilted plane wave at the two slits:

$$\theta = 2\pi\alpha\,\frac{d}{\lambda}\,, \tag{3}$$

whore d is the distance between the two slits. The intensity at point O becomes zero when the two waves are *out of phase, i.e.* $\theta = \pi$, which corresponds to the case, $\alpha = \lambda/2d$. This means that if uncertainty α in electron direction is larger than $\lambda/2d$, the interference fringes disappear by the overlap of variously displaced interference fringes. Consequently, if an interference pattern is to be formed between two slits separated by d, α must conform to the inequality,

$$\alpha < \frac{\lambda}{2d}\,. \tag{4}$$

In other words, an electron having uncertainty α in direction is *coherent* at two points separated by distance d only if $d < \lambda/2\alpha$. The maximum distance of the coherence is called the *spatial coherence length D* of the electron:

$$D = \frac{\lambda}{2\alpha}\,. \tag{5}$$

Since an interference pattern cannot be observed with a single electron, we have to use an electron beam consisting of successive electrons of similar nature to form an interference pattern by accumulating many electrons. If each electron has uncertainty in α in direction, the electron beam has the divergence angle α. We calculate how small α must be by using Equation (4). In a typical electron diffraction experiment, $d = 3$ Å and $\lambda = 0.04$ Å. Therefore the divergence angle of an electron beam should be less than 7×10^{-3} rad, or 0.4 degree. It is not difficult to obtain such a degree of the divergence angle of the electron beam. But when we want to get an interference pattern of two slits separated by a macroscopic distance, say, $d = 10$ μm, then $\alpha < 2 \times 10^{-7}$ rad $= 1.2 \times 10^{-5}$ degrees. We have to use a highly collimated electron beam.

There is one more condition for the formation of an interference pattern. The uncertainty in the speed of an electron has to be small. The wave corresponding to the electron consists of plane waves with different wavelengths. If the wavelength is halved, the fringe spacing is also halved. Therefore, interference fringes thus disappear by the overlap of fringes having various spacings when an electron has uncertainty in its speed.

An Electron Beam in an Electron Microscope

Macroscopic interference fringes have been observed for the first time by using electron microscopes, for which there was a reason. It is related to the situation that there are no convenient concave lenses in the world of electron microscopes. Since an electron lens inevitably has a large chromatic aberration, that is, the focal length of the lens is different for electrons with different speeds, only an electron beam having a narrow spread in the speed can be used. Furthermore, due to a large spherical aberration of an electron lens, only the central part of the lens can be used to get a sharp image, since the focal length of the peripheral part of the lens becomes shorter than that of the central part of the lens due to the spherical aberration. It was, therefore, necessary to develop beams that were uniform in speed and yet bright even when electrons were collimated to pass through the central part of the lens.

In 1940 macroscopic electron interference fringes were observed by H. Boersch in Germany. He observed *Fresnel fringes* like those shown in Fig. 18, which is, in a

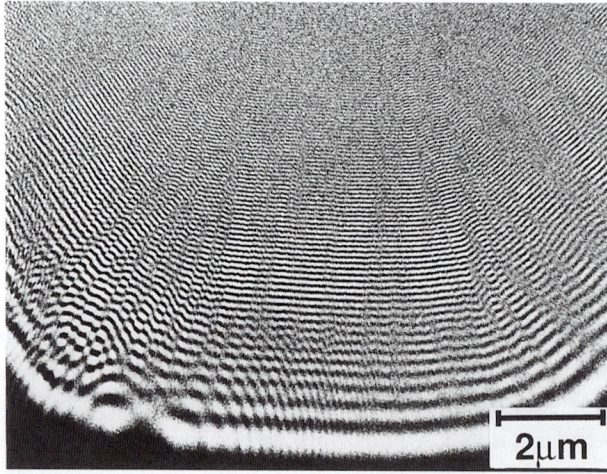

Fig. 18. Fresnel fringes.

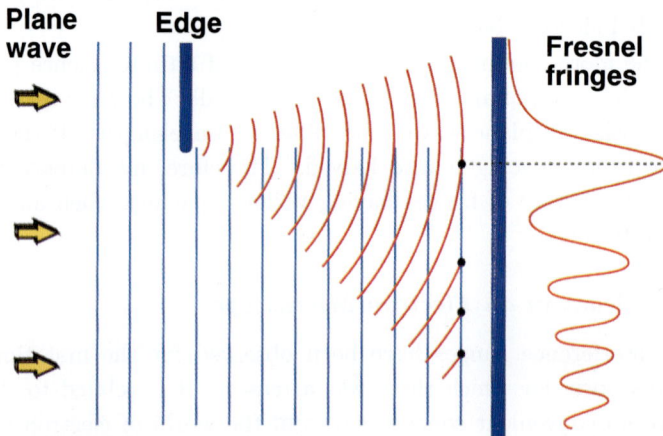

Fig. 19. Formation of Fresnel fringes.
    Fresnel fringes are formed between a plane wave and a cylindrical wave formed at the edge of the screen.

word, a greatly defocused image of a sample edge taken when an incident beam is collimated. As can be seen from the illustration in Fig. 19, these fringes are formed by the interference between the cylindrical wave scattered from the edge and a plane wave passing outside the edge. The fringe spacing becomes shorter with increasing distance from the edge. So many Fresnel fringes as are seen in Fig. 18 can only be obtained under the condition that a divergence angle of an incident electron beam is less than 1×10^{-6} rad.

One or two fringes can be observed even with a conventional electron microscope where the divergence angle is around 10^{-3} rad, or 0.06 degrees. Fresnel fringes produced along a specimen hole edge have been used for adjusting the astigmatism of the objective lens in electron microscopes.

Electron Biprism

In 1955 an electron interferometer using a device called an *electron biprism* was invented by G. Möllenstedt (Fig. 20) and H. Düker at Tübingen University in Germany. From the word *biprism*, you must not imagine an optical biprism made of two triangular-shaped glasses. Because electrons can only travel in vacuum, something different and yet that works like an optical biprism is necessary. A device with a structure shown in Fig. 21 was conceived to satisfy this condition. It is a thin filament lying in between two parallel-plate electrodes. Electrons go through both sides of the filament. When a positive voltage is applied to the filament, electrons, being negatively charged, are deflected toward the center. The two beams having passed through different sides of the filament overlap below. If the two beams are coherent, interference fringes are formed in the overlap region. With the biprism, all the electrons are deflected by the same angle irrespective of the position of an incident electron. Electrons passing near the filament are subjected to a stronger force and the time they spend there is short. On the other hand, electrons passing

Fig. 20. G. Möllenstedt (1912–1997).

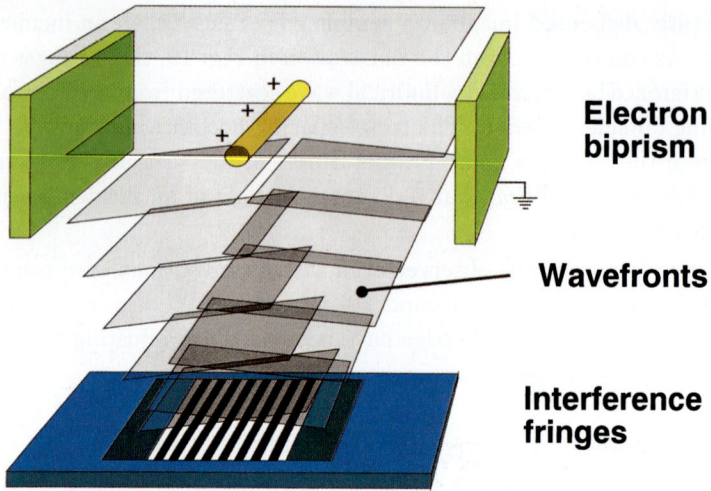

Electron biprism

Wavefronts

Interference fringes

Fig. 21. Electron biprism.

The electron biprism consists of a central thin filament and two parallel-plate electrodes. When a positive voltage is applied to the filament, electrons passing through on both sides of the filament are deflected towards the center. Since the deflection angle is the same irrespective of the incident positions, it acts as a biprism.

farther from the filament are subjected to a weaker force and for a longer time. The result is that all the electrons incident at any position are deflected by the same angle.

In this way, a plane wave incident onto the biprism is split into two, and two plane waves are tilted toward the center to produce interference fringes which are equally spaced as shown in Fig. 22. The angle of the wavefront-tilting and also the width of the interference pattern region are proportional to the voltage applied to the filament.

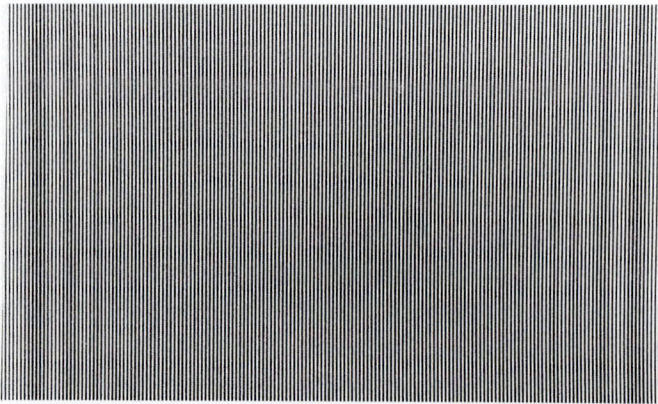

Fig. 22. Biprism interference fringes.

How to Fabricate the Filament Electrode

The separation between two electron beams must be less than 10 μm or so if they should interfere, and consequently the biprism filament must be much thinner than this. Making such a thin filament takes much care. One method is this: melt a glass rod with a burner and stretch it to both sides forcibly; then you will get an extremely thin, almost invisible thread. All the surfaces of the thread are then made to be conductive by covering them with an evaporated gold thin film. I can imagine how difficult it was for Möllenstedt to fabricate such a thin filament for the first time. He told me that spiders which produced uniform and thin filaments were searched for in a wood. When the spiders were kept for a while, the diameters of the filaments they spun changed when the season changed.

Using this electron biprism, we can see that electrons behave like waves, just like the string waves and water waves we saw in Figs. 8 and 13. Thus we can use an electron beam to take a hologram and then use a light beam to reconstruct an image.

Chapter 4

ELECTRON HOLOGRAPHY

You know about holography. Optical holograms are now used even in credit cards. You can see a three-dimensional image from a hologram. The principle behind this unique imaging was invented in 1948 by D. Gabor (Fig. 23), who aimed at improving the resolution of electron microscopes. Let me start by explaining the reason why holography had to be invented.

Fig. 23. D. Gabor (1900–1979).

Birth of Electron Microscopes

In the late 1920s, D. Gabor, born in Hungary, went to Germany to study at the Technical University of Berlin. There he investigated the cathode ray oscillograph. At first the cathode ray was not known to be a stream of electrons and got its name because it was produced from the cathode in a vacuum tube. Gabor used an electromagnetic coil to focus the cathode ray, a prototype of a magnetic electron lens. A few years later, H. Busch found that theoretically a cylindrical magnetic field acts as a lens for electrons.

Five years after Gabor left this university, an electron microscope was born there. The resolution of optical microscopes was known to be limited fundamentally by the wavelength of the light they used, and just at that time de Broglie's hypothesis that electrons have extremely short wavelengths began to be supported by experiments. The wavelength of yellow light, for example, is 6000 Å, while the wavelength of electrons accelerated up to 100 kV is only 0.04 Å. In the laboratory that Gabor left, E. Ruska was investigating electron lenses. All the necessary conditions were prepared for the birth of electron microscopes: conditions such as the discovery that electron waves have extremely short wavelengths, and the invention of electron lenses.

In 1932 Ruska and M. Knoll developed the electron microscope and demonstrated its principle. The resolution at that time did not even reach that of optical microscopes, but the resolution surpassed the optical resolution within a year. With electron microscopes it became possible for the first time to see such microscopic objects as viruses and crystal dislocations.

Aim of Gabor

Everybody thought that the resolution of electron microscopes would reach the fundamental limit — *i.e.* the electron wavelength — but an unexpected obstacle was soon revealed: the resolution was limited by the aberrations associated with an electron lens!

An optical lens can be made aberration-free by using a combined system of convex and concave lenses to compensate for aberrations [Fig. 24(a)]. While it was proven by O. Scherzer in Germany that there was no concave electron lens if the lens was made of an axial-symmetric magnetic field; the aberration compensation cannot be made by combining lenses in the case of electron lenses. With an ideal lens a point object should be focused to an image whose diameter is of the order of the wavelength, but it is instead greatly blurred because of the aberrations in the image-forming lens [Fig. 24(b)].

Invention of Holography

Many ideas to compensate for the aberrations were proposed in those days and the most unique was the *holography* Gabor devised. During the Easter of 1947 an unexpected idea flashed across Gabor's mind when he was waiting for his turn to play tennis. This idea was first called "microscopy by reconstructed wavefronts" and was later named "holography." The word originates from the Greek *holos* (whole) and *graphein* (to write) and it means photography with all the information covering both intensity and phase. An electron image is transformed into an optical image through holography. Then the effect of the aberrations in the image can

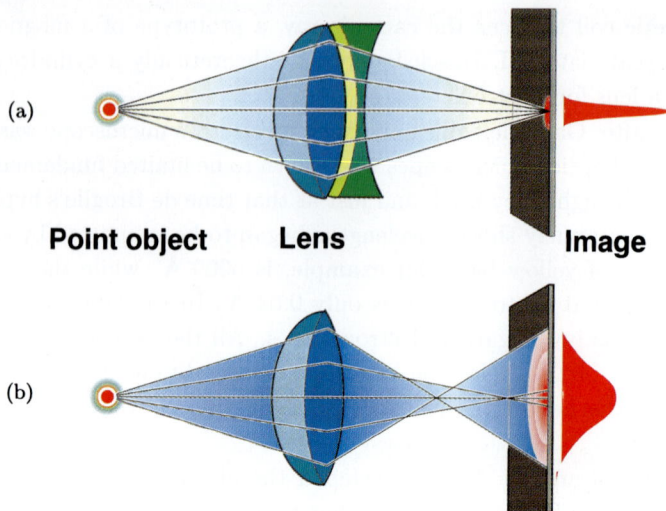

Fig. 24. Effect of lens aberrations.
(a) Optical lens
(b) Electron lens
Lens aberrations can be compensated for by combining concave and convex lenses as in case of an optical lens. While no concave lenses are available in electron microscopes; the image of a point object is blurred due to the aberrations.

be compensated for by using the opposite aberrations of an optical concave lens. With the improved resolution, Gabor wanted to see individual atoms.

Hologram Formation

Holography is a two-step imaging method: a special photograph called a *hologram* is taken with an electron beam, and then an optical image is reconstructed by illuminating the hologram with a light beam. Holography does not require any lens for image formation, but uses the fundamental properties of waves. We consider the principles of holography: how images are formed in two steps, hologram formation and image reconstruction, all without recourse to lenses.

First, I will explain the image formation by holography without using any mathematics. An object is illuminated by an electron plane wave [Fig. 25(a)]. For simplicity, a point object is selected as the object. If you want to get a concrete image of a plane wave you can imagine sea waves breaking on the shore, or recall a plane wave in the ripple tank in Chapter 2: the contours of the crests are wavefronts and a plane wave has planar wavefronts (linear in Fig. 12).

A plane wave collides with the point object O in Fig. 25(a), thus producing a spherical wave. The plane wave and the spherical wave proceed together, overlapping each other. Recall the interference experiment with two water waves (Fig. 14). The plane wave is the reference wave, and the spherical wave here is the object wave corresponding to the water wave disturbed by the object.

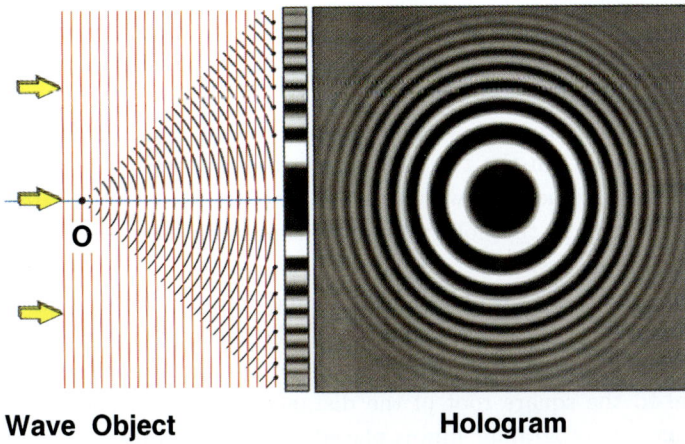

Wave   Object   Hologram

(a)

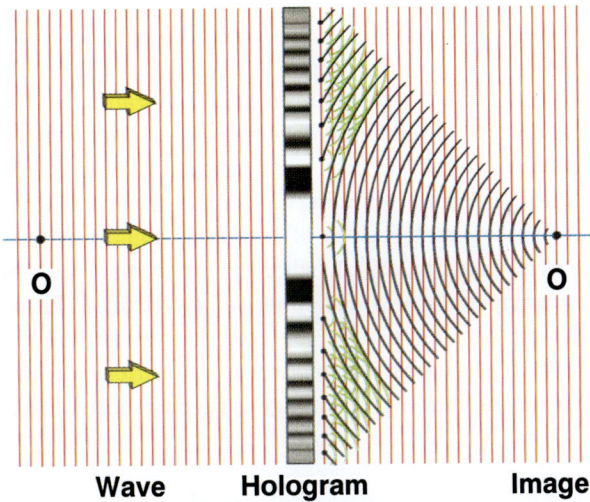

Wave   Hologram   Image

(b)

Fig. 25. Principle of holography.
(a) Hologram of a point object (zone plate)
(b) Image reconstruction from the reversed hologram

Let us consider the resultant intensity of the two waves in the hologram plane. The two wavefronts coincide at the central point in the hologram plane if the spherical wave produced is *in phase* with the plane wave as shown in Fig. 25(a). The crest of the plane wave meets the crest of the spherical wave to form a higher crest, just like the water surfaces of the two waves overlapping in phase and oscillating up and down. The oscillation of the electron wave is vigorous at this central point.

"How does the wave oscillate at other places in the hologram plane?" The phase difference of the two waves increases in proportion to the square of the distance from the central point in the hologram.

The oscillation disappears where the phase difference is π or the two waves are *out of phase*, since then the two waves cancel each other. At a point where the phase difference becomes 2π, two waves are again in phase to oscillate. In this way, the wave oscillation in the hologram plane repeatedly appears and disappears to form interference fringes.

In the present case of a point object, things remain the same if the system is rotated around the optical axis. Therefore the resultant interference fringes are concentric as shown in Fig. 25(a). The fringe spacing becomes shorter inversely proportional to the square root of the distance d from the center of the circular fringes. When a photographic film is placed on the hologram plane and is exposed, the film is blackened in proportion to the intensity of the oscillation of the electron wave, that is, in proportion to the square of the amplitude. This hologram of a point object is called a *zone plate*.

Image Reconstruction

An optical plane wave is then incident onto this hologram film [Fig. 25(b)]. In this figure, the black and white of the film is reversed for simplicity. Light passes through only the transparent parts of the film. The overall wavefronts of the transmitted light can easily be obtained by summing up all the spherical wavelets produced at every point in the transparent regions of the film. Some wavefronts of the spherical wavelets are drawn in Fig. 25(b), where you can see that new wavefronts are produced as the envelopes of the many spherical wavefronts. These new wavefronts are concentric spherical surfaces with the common center at O'. The light wave transmitted through the hologram reconstructs a spherical wave converging to single point O'. This means that a point image is reconstructed from the hologram of a point object.

Restrictions of Objects in In-line Holography

Since an arbitrary object can be regarded as consisting of point objects, the object should be reconstructed from its hologram. However, an arbitrary object cannot really be reconstructed in this type of holography. To see why, we have to understand the properties of a zone plate in more detail.

If you consider carefully the wavefront reconstruction in Fig. 25(b), you will realize that another kind of wavefronts are produced as the envelope of the many spherical wavelets. The new wave is a spherical wave diverging from point O. After all, a zone plate has three kinds of functions for an incident light plane wave: that of a parallel transparent plate through which the plane wave passes without any influence, that of a convex lens through which the plane wave is converted into a spherical wave converging at O', and that of a concave lens through which the plane wave is converted into a spherical wave diverging from O.

This method of holography that Gabor devised is called *in-line holography*, since an object wave (spherical wave) and a reference wave (plane wave) propagate in the same direction. As you may have noticed, an object has to be small with this method. This is because a large object can disturb the plane wave, which can then no more be regarded as a reference plane wave, and also because the two images formed at O and O' disturb each other.

Early Experiments

Gabor first attempted to confirm his idea with experiments using light, but in those days it was not easy to obtain coherent light beams. A spectral line from a high-pressure mercury lamp was used for the light source, and a point source was obtained

(a)

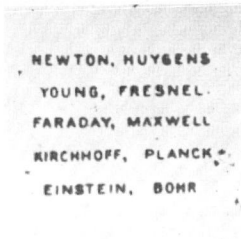

(b)

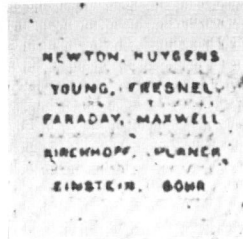

(c)

Fig. 26. Experimental results from optical holography using light obtained by Gabor (D. Gabor: Proc. R. Soc. London **A197** (1949) 454).
(a) Hologram
(b) Object
(c) Reconstructed image

by using a small pinhole of a diameter of 2 μm (2×10^{-4} centimeter). However, no matter how hard one strained his eyes in the dark, no interference fringes could be seen with the naked eye. And with a fast film for x-ray and an exposure time of several minutes, two hundred interference fringes were barely recorded.

In less than one year, however, Gabor succeeded in obtaining reconstructed images (Fig. 26). The hologram looked like just the ripples on the water surface [Fig. 26(a)], yet the letters of "NEWTON," "HUYGENS," and other names were clearly recognized in his reconstructed image [see Fig. 26(c)]. Gabor's theory was confirmed by his own experiment.

Gabor then resolved to try holography using an electron beam. In 1948 he was with BTH (British Thomson Houston Co.) in Great Britain. Its parent company AEI (Associated Electrical Industries) built a new research laboratory, and research on electron microscopes was started as one of their long-term research projects. Gabor proposed to join the project to do electron–holography research, but BTH refused to let him go because of some contractual matter. So the project ended up being carried out by a team led by M. E. Haine and T. Mulvey at AEI under the direction of Gabor. The research funds for this work were for the first time given to a private company by the government. After several years, they obtained reconstructed images from Fresnel-fringe-like holograms (Fig. 27). But they reached the conclusion that holography, which was difficult to realize with light, was even more troublesome with electrons and so had to be abandoned. In 1948 Gabor was invited to Imperial College in London and left BTH. As time passed, his interest in holography — and that of the public — seemed to fade away.

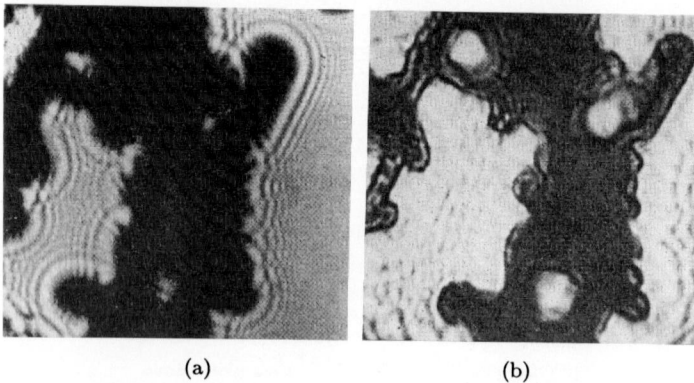

(a)   (b)

Fig. 27. Experimental results from electron holography obtained by Haine and Mulvey (M. E. Haine and T. Mulvey: J. Opt. Soc. Amer. **42** (1952) 763).
(a) Electron hologram
(b) Optically reconstructed image

In Japan, T. Hibi in Tohoku University was interested in holography and performed experiments with an electron beam. He developed a new electron gun because he recognized that making an electron hologram would require a coherent electron beam. The conventional electron gun used thermal electrons emitted from a heated hairpin-shaped tungsten filament one tenth of a millimeter thick, whereas Hibi's new electron gun had a needle at the tip of the hair pin and emitted an electron beam from the needle tip. This *pointed filament* is still used for high-resolution electron microscopy.

Emergence of Laser Holography

The great leap forward in the history of holography happened in 1960. Until then holography was recognized to be interesting in its principle, but it remained impractical owing to the difficulty of obtaining coherent waves. E. N. Leith and J. Upatnieks of the United States, however, got beautiful images by taking advantage of the newly born *lasers*. They did not just follow the in-line method that Gabor proposed as explained in Fig. 25. They adopted a new method called *off-axis method* which resulted in clear reconstructed images (see Fig. 28). In this method, an object wave and a reference wave do not proceed in the same direction, but at an angle to one another. As already explained, an in-plane hologram of a point object is a zone plate which consists of concentric circular fringes as illustrated in region A in Fig. 29. While an off-axis hologram of a point object can be considered to be a peripheral part of a zone plate which consists of nearly parallel fringes as illustrated in region B in Fig. 29, illuminating this "grating" with a plane wave therefore produces two diffracted waves in each of which a reconstructed image and a conjugate image are reconstructed.

The reason for the image formation is as follows. An in-line hologram of a point object, a zone plate, has the functions of convex and concave lenses as can be seen

Fig. 28. Reconstructed image from an off-axis hologram obtained by Leith and Upatnieks (E. N. Leith and J. Upatnieks: J. Opt. Soc. Amer. **57** (1967) 975).

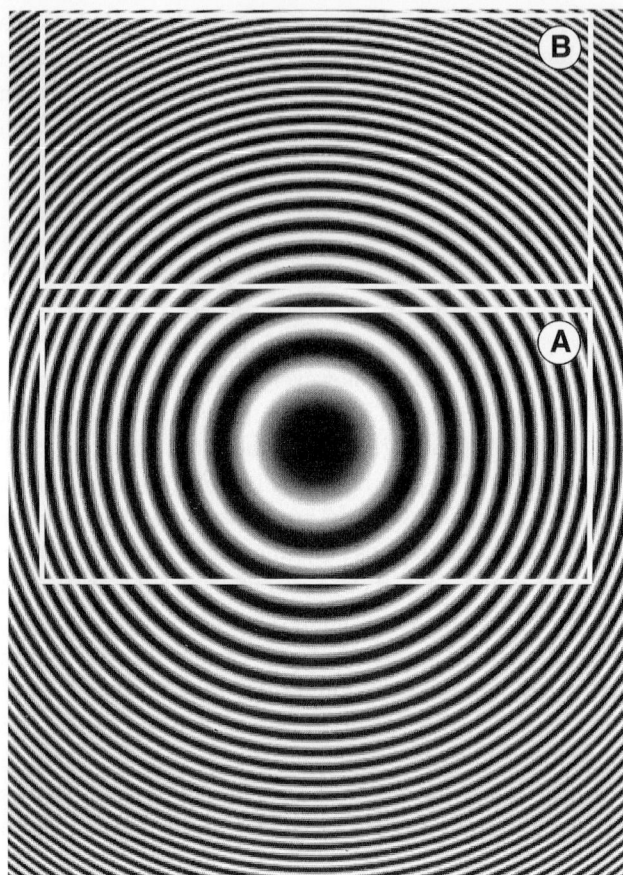

Fig. 29. In-line and off-axis holograms.
  The central part (A) of a *zone plate* is an in-line hologram of a point object and the
  peripheral part (B) of it is an off-axis hologram of a point object.

in Fig. 30(a) and (b). For an incident plane wave, this zone plate has real and
virtual focal points at O' and O. Since an off-axis hologram of a point object is also
a zone plate but in a region far from the center, this hologram has the functions of
the peripheral parts of convex and concave lenses as shown in Fig. 30(c) and (d).
The same thing happens when a plane wave is incident on an off-axis hologram,
or peripheral parts of a lens: the real image is formed at O' and the imaginary
image is formed at O. The two image-forming waves are both off-axis propagating
in different directions, and therefore the two images can be observed separately. The
phase distribution of the image formed at O is the same as that of the original object,
whereas the phase distribution of the other image at O' is the same in absolute value
but reversed in sign, or complex-conjugate. It is therefore called a *conjugate* image.
Optical holography was developed rapidly by using both a coherent laser light and
the off-axis method, and it began to be applied in numerous fields.

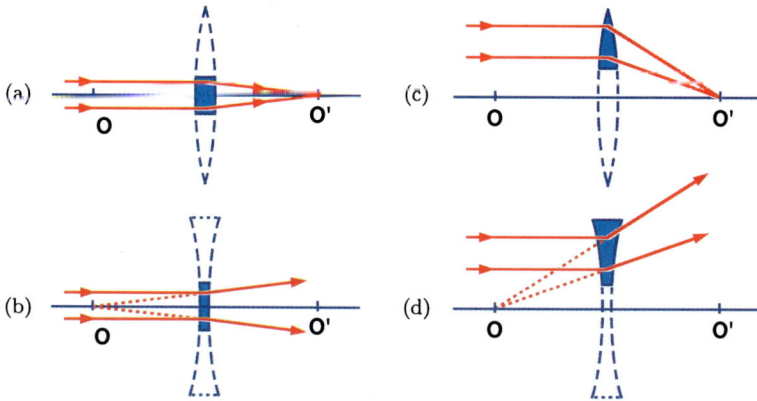

Fig. 30. Functions of a zone-plate hologram.
    (a) Convex-lens function of the in-line hologram
    (b) Concave-lens function of the in-line hologram
    (c) Convex-lens function of the off-axis hologram
    (d) Concave-lens function of the off-axis hologram

Holography Became Feasible with Electrons!

Stimulated by the new developments of optical holography, my colleagues and I at Hitachi Central Research Laboratory devoted ourselves to the realization of electron holography. The fact that a laser was now available to us did not allow us to be optimistic. The problem lay not in image reconstruction with a laser, but in hologram formation with electrons. Therefore, we made efforts to form higher-quality electron holograms. First, we used the point filament developed by Hibi, sacrificing filament lifetime to obtain a brighter electron beam. Next, we dared to increase the exposure time. Electron micrographs are generally taken within a few seconds, and it was extremely difficult to fix a sample for several minutes. It was fortunate that we were able to use the high-performance microscope with which T. Komoda established the world record in resolving power.

Furthermore, we adopted a new condition in making holograms. In the in-line method, when one tries to see a reconstructed image, the other conjugate image inevitably overlaps to disturb it. But with the off-axis method proposed by Leith and Upatnieks, twin images are formed in the two beams traveling in different directions. Thus each image can be observed independently. We knew that the off-axis method was more convenient for obtaining disturbance-free images, but did not use it because of its strict requirements for electron beam coherence.

At about the same time, B. J. Thompson of the United States proved in optical holography that even in Gabor's in-line method the disturbance of a reconstructed image by its conjugate image could be made negligible. In this type of holography called Fraunhofer holography, the locations of two reconstructed images were apart by a large distance. Therefore the conjugate image was completely blurred and becomes a constant background in the plane of the reconstructed image, and the reconstructed image could thus be observed without any disturbance. We started to make experiments but it turned out that they were very difficult. For example, to take one hologram, we had to hold with bated breath for more than ten minutes in the dark and to keep absolutely still.

Such experiments went on for about a year, and in 1967 we finally succeeded in getting a clear image not affected by its conjugate image. An example of the results obtained at that time are shown in Fig. 31, where the specimen consists of fine gold

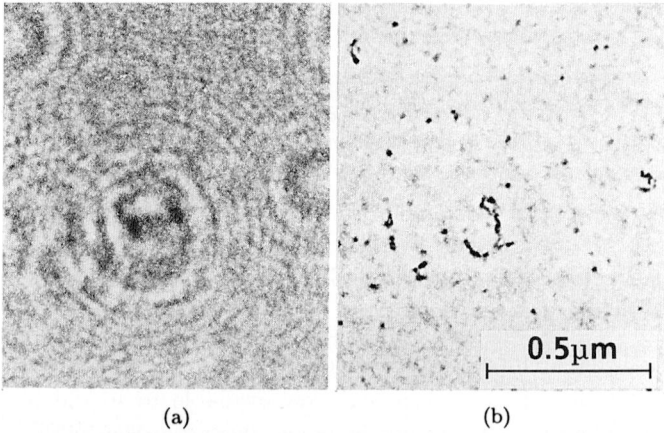

(a)   (b)

Fig. 31. Image reconstruction from an in-line electron hologram of fine gold particles.
(a) Hologram
(b) Reconstructed image

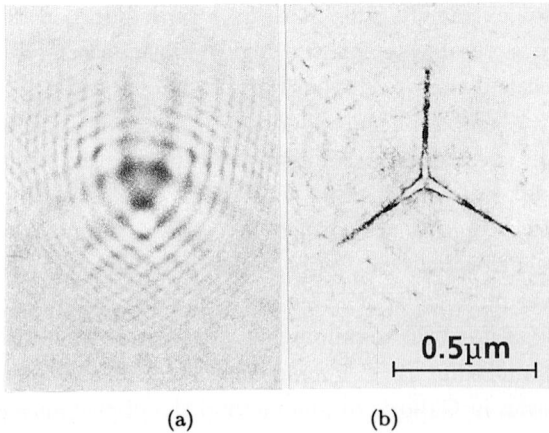

(a)   (b)

Fig. 32. Image reconstruction from an in-line hologram of a zinc oxide smoke particle.
(a) In-line hologram
(b) Reconstructed image

particles. One can see a little-modulated zone plate in the center of the hologram shown in Fig. 31(a). From the zone plate, the image of an aggregate of gold particles is optically reconstructed as can be seen from Fig. 31(b). It is interesting to note that even from the peripheral region of the hologram where nothing like zone plates can be recognized, individual particle images are reconstructed.

Another example is shown in Fig. 32, where the specimen is a zinc oxide smoke particle. The particle is shaped like a tetrapod and consists of four thin needles. The hologram does not look like a zone plate at all. A needle can be considered to consist of points in a line and consequently its hologram is the sum of zone plates parallel-transported in the direction of the needle resulting in parallel interference fringes. It is a one-dimensional zone plate. One-dimensional zone plates oriented in three directions overlap to form the hologram shown in Fig. 32(a).

These needles are reconstructed in Fig. 32(b). We can optically focus the image at the tip of any of the needles. When looking at these photographs, I cannot help but feel the marvel of holography. Sometime later, using an electron biprism, Möllenstedt and H. Wahl in Tübingen University, succeeded in obtaining an image from the off-axis hologram (Fig. 33).

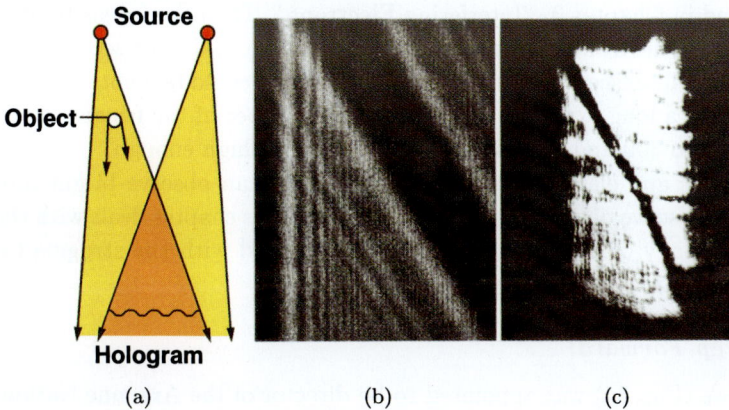

(a)   (b)   (c)

Fig. 33. Image reconstruction from an off-axis hologram of a filament (G. Möllenstedt and H. Wahl: Naturwiss. **55** (1968) 340).
(a) Experimental arrangement
(b) Off-axis electron hologram
(c) Reconstructed image

Thus, electron holography was verified experimentally. We also attempted, as Gabor had, to improve the resolution of the electron microscope. It was soon recognized, however, that the coherence of an electron beam was so poor that the highest resolution of reconstructed images attainable could not reach even the resolution limit of the current electron microscopes. The real advantages of electron holography cannot be manifested without coherent electron beams comparable to optical lasers. With this strong belief, we started to develop a coherent electron beam.

Chapter 5

COHERENT ELECTRON BEAMS
DEVELOPED!

Several kinds of electron interference fringes have been introduced up to now: interference patterns of electron waves scattered from single crystals (electron diffraction patterns), Fresnel fringes, biprism fringes, and holograms. These interference fringes were, however, difficult to obtain before 1978, though the coherence of an electron beam had already been improved to some extent compared to that in the days of Gabor. In order to form interference fringes, an electron beam has to be collimated as explained in Chapter 3, "Interfering Electrons." Then the intensity of a conventional thermal electron beam became so low that we could not see the interference fringes on the fluorescent screen directly. The fringes could be photographed on film, but only after a long exposure time. The total number of the fringes photographed were limited to 300 and the fringe contrast was not high enough.

"Are there any electron beams with which we can observe bright interference fringes like those we observe with a laser beam?" This chapter deals with the history of the technology of the coherent electron beam and with the struggle to develop this kind of beam.

A Big Step Forward!

A. V. Crewe (Fig. 34) was appointed to be director of the Argonne National Laboratory, which had a staff of five thousand, when he was only thirty-four. After six years' tenure, he left his administrative position in order to pursue his ambition to observe single atoms. He returned to the University of Chicago as a scientist so that he could put the field-emission electron gun to practical use. That was a dream of many people in the 1950s, but it had not been realized. Without Crewe's success there would have been no more progress in electron holography. Now let us begin with the story of this *dream electron gun.*

What is a Coherent Beam?

First of all, we have to know what kind of properties a coherent electron beam has to have. As we touched on in Chapter 3, an electron beam has to be parallel and monochromatic to form interference fringes. Therefore, we need a new kind

Fig. 34. A. V. Crewe (1927–   ).

of electron beam, one that is bright even when the electron beam is made to be both parallel and monochromatic. Of these two features, the beam parallelism is more important than the beam monochromaticity for the practical observation of electron interference fringes. For example, if you want to get a much larger number of interference fringes than 100, the divergence angle of a 100 kV electron beam has to be made smaller than 1×10^{-5} rad but only at the cost of the current density of the beam. The current density of a beam cannot be increased by any electron-optical arrangement when the divergence angle is fixed, but is determined by the property of the beam called the "brightness." If the beam brightness increased 100-fold, the interference patterns would become 100 times brighter.

"What is the brightness of an electron beam?" The brightness R is the electron current density i per unit solid angle Ω:

$$R = i/\Omega. \qquad (6)$$

The solid angle of the beam is, so to speak, a two-dimensional divergence angle of the beam. We already used the divergence angle α of the beam in Equation (4) in Chapter 3, but that was the angle in a plane. The direction of electron velocities in the beam are not confined in a plane, but are also scattered in the direction perpendicular to the plane. When the distribution of the divergence angle is axially symmetric, the solid angle Ω can be defined as $\pi \alpha^2$ when α is small. Here it may be better to specify the value of α more exactly: Let us consider such electrons in an electron beam that pass through a point in the beam. Then α is given by the width of the angular distribution of the directions of these electrons.

There is a theorem that this value of R, the current density per unit solid angle, remains the same at any cross-section of the beam, even when lenses or pin holes are inserted in the beam path. Therefore, we cannot increase the value of the beam brightness by any electron-optical means. We have no choice but to use a new kind of bright electron source to get a brighter beam.

Bright Electron Beam

"How can we get an electron beam having a high brightness?" As can be seen from Equation (6), we can get a high brightness by making the current density i large or by making the solid angle Ω small.

To increase the current density of thermal electrons at the cathode, the temperature of the cathode is raised to energize electrons so that many electrons can get over the surface barrier (explained a little later) and come out of the cathode. The temperature we can use is limited, however, by the melting or evaporation of the cathode material.

"Are there any methods to extract electrons without heating the cathode?" Yes, there are several methods, and in particular the current density of *field-emission electrons* is conspicuously high.

Electrons Emerging by "Tunnel Effect"

The phenomenon that electrons are emitted in response to the application of a strong electric field at a metal surface was discovered in the 19th century but could not be explained in those days.

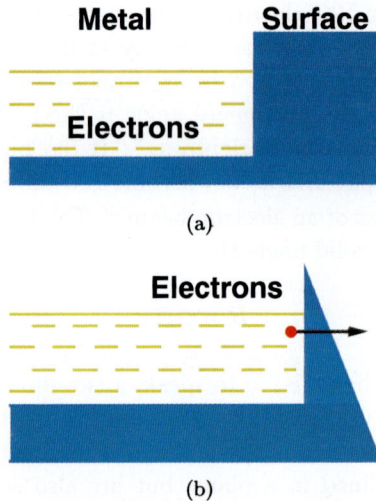

(a)

(b)

Fig. 35. Electron emission from metal.
    (a) Electrons confined inside metal by a surface barrier
    (b) Emission of electrons by tunneling through a thin surface barrier

Electrons exist inside atoms as we've already studied. When atoms get together to form a metal, the outer electrons surrounding the nuclei can freely move inside the metal. It is these *free electrons* that make it possible for an electric current to flow through a metal. Electrons cannot go outside the metal, however. A surface barrier is produced so that electrons cannot flow outside the metal [Fig. 35(a)]. The electrons are just like water in a pool.

"What happens when a strong electric field is applied to the metal surface?" Nothing happens inside the pool, but only the shape of the barrier changes. The ground level outside the pool is tilted and the wall at the surface becomes thin [Fig. 35(b)]. According to classical mechanics, no matter how thin the barrier is, an electron running against the barrier is always bounced back by it and can never pass through it. According to quantum mechanics, however, the situation is entirely different. The uncertainty principle tells us that we have no means to confirm whether or not an electron is bounced back by an extremely thin barrier: when the barrier is only a few angstroms thick, the electron has uncertainty in its velocity near the barrier and there is some probability that the electron can penetrate the barrier, which is called *field emission.*

Our question then is how we can induce this field emission. The naive tactics would be to apply a high voltage between two parallel metal plates. Probably 100,000 volts could be applied between two polished plates 1 cm apart. The thickness of the wall then is only a few thousand angstrom, but this is still not thin enough. A thousand times higher electric field is required to make the wall a few angstrom thick for inducing electron emission.

Electrons Emerging from a Needle Tip

Now think about a lightning rod like that illustrated in Fig. 36. Lightning does not strike the open field, but the lightning rod. The reason is that electric lines of force tend to gather at a sharp tip. The lightning rod has a sharp tip so that the lines of force from the lightning clouds are concentrated at the tip of the rod, and the strong electric field attracts lightning. Learning from this, we use a piece of wire which has a thickness about that of a human hair, 0.1 mm, and has a pointed tip at one end (Fig. 37).

You may wonder how one can fabricate such a fine needle. It is not so difficult. You can fabricate such a tip simply by dipping a tungsten wire into a solution of calcined soda and then applying an electric current between the wire and an electrode. The wire is etched selectively near the solution surface and is eventually cut into two parts. When the current stops at that instant, a sharp tip is obtained. The other part of the wire has a similar tip, but drops into the solution. As long as the solution surface can be kept from vibrating, this beautifully shaped tip is formed.

How can Electrons be Emitted?

Electrons are emitted from the tip by the application of a high voltage between the tip and a nearby electrode so that the tip is negatively charged. If the tip is as sharp as 0.1 μm in radius, like the one shown in Fig. 37, it emits electrons at about a few thousand volts. Things do not go well, though, unless the tip is placed in an ultra high vacuum.

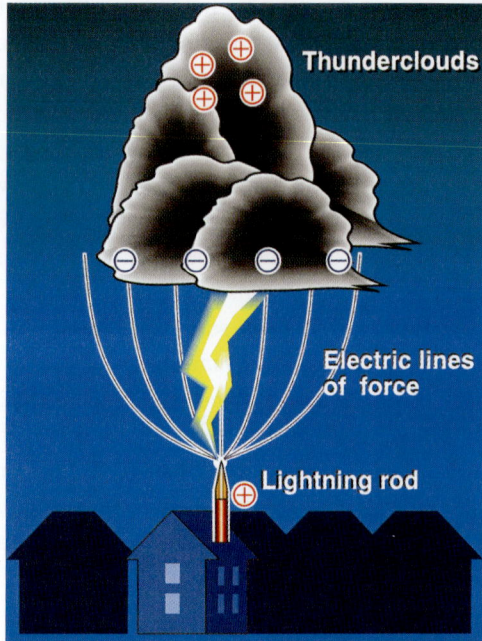

Fig. 36. Lightning rod.
Electric lines of force are concentrated on the tip of the rod.

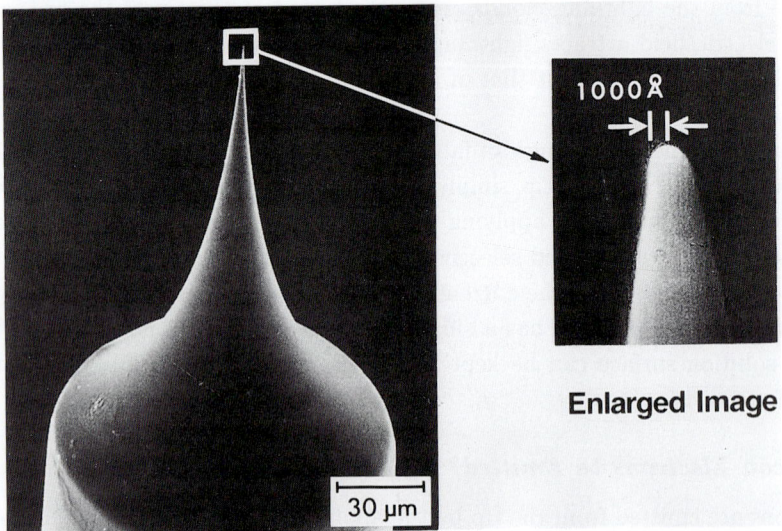

Fig. 37. Scanning electron micrographs of a field-emission tip.

I myself tried to get electrons in an ordinary vacuum, with a pressure of about 10^{-4} Pa. In an ordinary vacuum, however, the emission stops instantly. When we then look at the tip, we find it is no longer sharp, but rounded. The tip melts because of the momentary discharge caused by the residual gases still in the vacuum. Such a discharge ruins the tip, so we must have a nearly perfect vacuum like that in outer space to get stable electron emission. The pressure in the vacuum should be less than 10^{-7} Pa.

But troubles come not only from the vacuum. When the tip vibrates mechanically or the electron beam emitted from it is deflected by stray magnetic fields, the coherence of the beam is soon degraded. Because the source size is as small as 50 Å, a source movement of only 150 Å decreases the brightness by an order of magnitude.

Crewe was the person who made such a troublesome device into something usable. I can imagine how difficult it was; when I was trying to find out the optimum conditions following his path, I myself often wondered, "Is this really usable?". When I think of Crewe who hewed his way through an untrodden land, I realize that his greatest virtues were his strong belief and his courage. With this electron beam, he got a bright and fine electron probe, and he became the first man who saw the clear image of single atoms.

Electron Beam Like a Laser

Encouraged by the success of our electron holography experiment in 1967, I was eager to make Gabor's dream come true. Through long, demanding experiments, I came to appreciate the necessity of having a coherent electron beam like a laser. It was then that I heard about Crewe's accomplishment. The news that Crewe had actually put the electrons emitted from a tip to practical use surprised everyone. During the 1950s people at several laboratories tried to use this field-emission electron beam for practical purposes. Even after ten years had passed, no success was reported, so some people at first did not believe that Crewe had actually succeeded. But gradually the value of his electron beam came to be appreciated. In fact, the resolution of scanning electron microscopes improved by one order of magnitude with his new electron gun, and his newly developed scanning transmission electron microscope made it possible to observe heavy atoms.

Needless to say, we adopted this electron gun. And not only us. Several projects all over the world started to use this electron gun as a coherent electron source. It took ten years for a field-emission electron beam to exhibit its full ability.

At first I didn't think it would take such a long time. Crewe made a bright point source. Once the source is located at the focal plane of a lens, a parallel electron beam should be obtained. Although Crewe solved the essential technical problems, a field-emission electron gun still had some other problems preventing its use as an electron source for coherent illumination. Rather than forming a fine probe on a specimen in case of a scanning microscope, a parallel electron beam has to illuminate a wide field of view of the specimen to form electron holograms. This requires a large current. Moreover, since holograms are formed using electrons transmitted coherently through a specimen, we need an electron beam having a high energy and hence a high accelerating voltage, at least, 100,000 volts.

In 1973, we developed an electron microscope equipped with a field-emission electron gun. The coherence of an electron beam was definitely improved but was not enough. We learned from the development that an initially coherent electron beam would be disturbed before it reached a specimen. Since the beam was very fine and parallel, it was influenced by the presence of very weak stray magnetic fields. We had to treat an electron beam in a completely different manner from that in a conventional electron microscope because of the following big difference in electron source size. The size of a conventional electron source is 10 μm, while that of our source is as small as 50 Å or 0.005 μm.

Our second electron microscope was completed in 1976. The budget for this microscope was provided by H. Watanabe, general manager of the Hitachi Central Research Laboratory, in 1974 when I came back from Tübingen after one year's study under Prof. Möllenstedt. The microscope constructed was not promising at first in spite of our careful and elaborate design by making the best use of the ealier experience. It was surely better than the previous one, but it still had problems. This time, we were pressed to succeed in the project, because we knew that this was the last chance given to us. For the next two years, we struggled to improve the beam coherence. We were fortunate to continue such a long-range research on electron holography in an industrial research laboratory, and the support we received came not only from Watanabe, but also from many other managers and administrators in Hitachi who admired and encouraged fundamental research even in basic sciences. It is said that when N. Odaira, the founder of Hitachi, established the Central Research Laboratory in 1942, he dreamt that some day Nobel prize winners would be born from this laboratory. T. Kanai, the president of Hitachi was once a researcher in the Central Research Laboratory.

On the Top of the Tower

We tried every means we could think of in our efforts to obtain a coherent beam. One episode is still vivid in my mind. We thought that we had removed all sources of disturbance that would affect the electron beam, such as stray magnetic fields coming from not only the computers in the next room but also the power supply of the microscope itself. But the beam disturbance was still there: when the beam was focused on the fluorescent screen, the small focused spot moved around on the screen drawing Lissajous figures. We wondered whether it might possibly be caused by trains passing near our laboratory.

One of us went up to the tower on the twelfth floor of the Central Research Laboratory (Fig. 38) and by phone he reported the coming-and-going of trains. We did not find any relationship between trains and the disturbance when the train

Fig. 38. Hitachi Central Research Laboratory, Kokubunji, Tokyo.

traffic was heavy in the daytime. Around midnight, as the number of passing trains decreased, the disturbance of the electron beam also decreased. We kept this procedure, and little by little we started to understand the influence of the trains on an electron beam. As soon as the last train passed by, the main disturbance vanished. The disturbance occurred when a train started at the Kokubunji station. Finding a cause is the same as solving the problem. One gives up when he does not know where to go and thus thinks that is the end of his advancement.

A few months later, we succeeded in having an excellent outcome with great excitement.

The Day the Beam Coherence was Improved!

Suddenly came the day that brought a dramatic improvement to the performance of our machine. In using electron microscopes, we usually observe images directly on the fluorescent screen. Observing interference fringes, however, was like photographing faint stars at night. Since no interference fringes could be observed directly on the fluorescent screen, we could not even tell whether the fringes were recorded until we developed the film. Even when they were, the number of fringes recorded was at most 300.

The corresponding voltage applied to the biprism filament was around 50 volts. Even if a higher voltage was applied in order to get more interference fringes, no fringes could be recorded because the fringes became too faint. While the situation was different with the newly developed machine, interference fringes could be still photographed even when the voltage surpassed 100 volts. Many batteries were provided for the experiment and a few hundred volts were applied to the filament as a trial.

It was at the end of 1977. T. Matsuda, who was photographing interference fringes, brought me a surprising film just developed.

We usually observe electron microscope films with the naked eye since the film area is ten times larger than that of the 35 mm film. Even when we want to examine the film in detail, it is sufficient to look at it with a magnifier. We could see nothing in the film on which Matsuda should have photographed the interference fringes. But something new was photographed on it, because when we held the film against the fluorescent lamp, the lamp was seen to have split into three images in rainbow colors. Fine gratings were recorded on the film!

I was surprised and excited when I examined the film with an optical microscope and saw an uncountable number of interference fringes. Until that time, I believed that the narrowest fringe spacing that could be recorded on electron microscope film was 20 μm or a little less, but this time interference fringes with 7 μm spacing were recorded on film. The number of fringes exceeded 3,000. This means that the brightness of an electron beam was improved by two orders of magnitude. And as a result, the quality of the holographically reconstructed images became as good as that of the electron microscopic images.

The story of the development of a coherent electron beam did not end here but is still going on. The obtained coherence of a field-emission electron beam is not high enough and it's still a long way to go before attaining that of a laser beam. We are making efforts to further improve the beam coherence.

Chapter 6

WAVE-PARTICLE DUALITY

Up to here, when talking about electron interference phenomena, I have been taking it as a matter of course that electrons are waves. Since we have seen several kinds of interference patterns, you may have been convinced that electrons are waves like the water waves seen in Chapter 2. It also remains true, however, that electrons are particles. An electron is always a particle when detected and has never been divided into two or more pieces. You may think it is paradoxical that electrons have both wave nature and particle nature, but it is the most fundamental principle of quantum mechanics, the law of the microscopic world.

Waves or Particles?

The particle nature means that an electron is localized at a point, and the wave nature means that an electron is extended in a space. "How can electrons have these apparently contradictory properties?" The *two-slit* experiment demonstrating the essence of this problem is almost always introduced at the beginning of quantum mechanics textbooks. I would like to discuss this most important experiment in quantum mechanics here, though many of you may already know about it.

The two-slit experiment is an interference experiment in which electrons or photons are incident one by one onto two slits (Fig. 39). The time interval between them is so sparse that at most one electron or photon exists in the apparatus. The experiment using photons, in place of electrons, is described in S. Tomonaga's essay "Trial of a Photon." In this essay, a prosecutor asserts that Miss Photon must have passed through one of the two slits since she has never been found split into two or more pieces. Note that a photon in Chinese characters can also be pronounced as Mitsuko, a girl's name. While she insists that the prosecutor is not right, her behavior is inspected at the scene of the event. We send out photons one by one onto the two slits. The arrival of individual photons is recorded on the screen behind the two slits. When the arrival spots are accumulated, their distribution shows evidence that photons did not pass through one of the two slits: The evidence is the interference pattern which is the result of the two waves having passed through the two slits. We have to accept Miss Photon's assertion that she has not passed through one of the two slits. Since the story of this trial is very interesting, I recommend

45

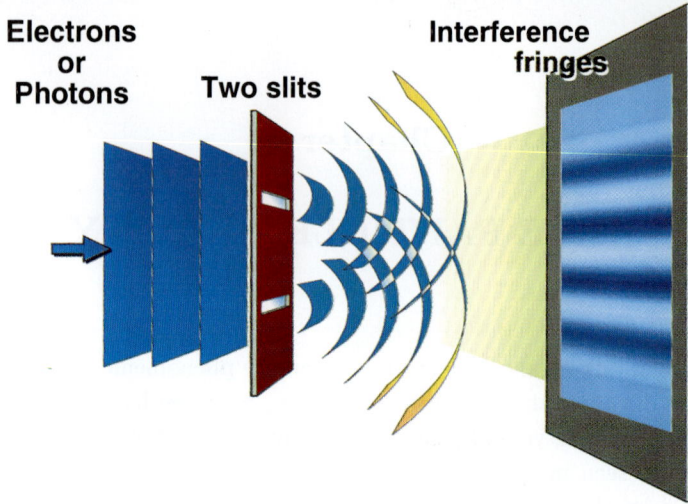

Fig. 39. Two-slit experiment.

that you read the original if you are proficient in Japanese: S. Tomonaga, *Quantum Mechanics and I*, ed. H. Ezawa (Iwanami, Tokyo, 1997).

To explain the significance of the two-slit experiment using electrons, I have no better words than those of R. Feynman, who shared a Nobel Prize with Tomonaga in 1965. At the beginning of the chapter on quantum mechanics in his textbook, *Feynman Lectures on Physics* (R. P. Feynman, R. B. Leighton and M. Sands: Addison-Wesley, Reading, 1965, Vol. III, Chap. 1), he writes as follows.

> In this chapter we shall tackle immediately the basic element of the mysterious behavior in its most strange form. We choose to examine a phenomenon which is impossible, *absolutely* impossible, to explain in any classical way, and which has in it the heart of quantum mechanics. In reality, it contains the *only* mystery. We cannot make the mystery go away by "explaining" how it works. We will just *tell* you how it works. In telling you how it works we will have told you about the basic peculiarities of all quantum mechanics . . .

> We should say right away that you should not try to set up this experiment. This experiment has never been done in just this way. The trouble is that the apparatus would have to be made on an impossibly small scale to show the effects we are interested in. We are doing a "thought experiment," which we have chosen because it is easy to think about.

Two-Slit Experiment with Electrons

This *thought experiment* has become feasible, however, thanks to the development of an electron microscope magnifying the microscopic world, an electron biprism forming a two-beam interference pattern, a coherent electron beam, and a

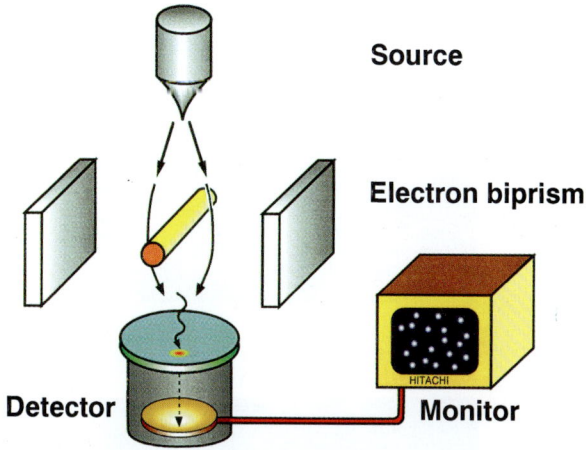

Fig. 40. Experimental arrangement of a two-slit experiment for electrons (A. Tonomura *et al.*: Amer. J. Phys. **57** (1989) 117).

two-dimensional electron counting system detecting electrons one by one with almost 100% efficiency.

Electrons emitted from the source pass through the electron biprism and enter the detector (see Fig. 40). When an electron arrives at the detector, a bright spot appears on the monitor. That is the place where an electron has arrived. Electrons are gradually accumulated on the monitor.

Let us start the experiment by emitting electrons one by one. When you look at the monitor, you can see that electrons begin to be detected here and there. Six frames of a videotape made at the very beginning of this experiment are shown in Fig. 41. You can see how electrons are accumulated one by one. Judging from these pictures, electrons are surely detected not as waves but as particles. Their arrival positions look random.

(a)

Fig. 41. Electrons detected one by one displayed on the monitor.
Number of electrons detected: (a) 1, (b) 3, (c) 5, (d) 8, (e) 14, (f) 19.

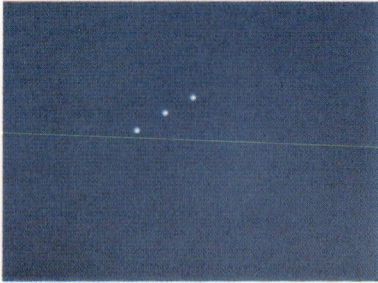

(b)

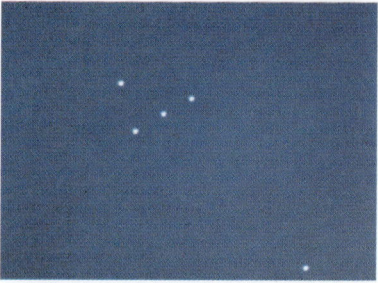

(c)

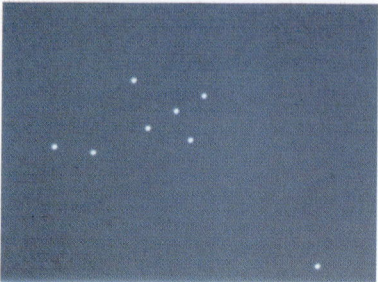

(d)

(e)

Fig. 41. (*Continued*)

(f)

Fig. 41. (*Continued*)

Electrons are arriving at a rate of several electrons per second in the field of view shown in Fig. 41. The electrons in this experiment are accelerated to 50,000 volts, and consequently the electron velocity is 120,000 km/sec, two fifth the speed of light. Electrons spend only 10 nanoseconds, *i.e.* 10^{-8} seconds in traveling the 1-m-long apparatus. There is no problem in supposing that each electron is detected in an instant after it is emitted.

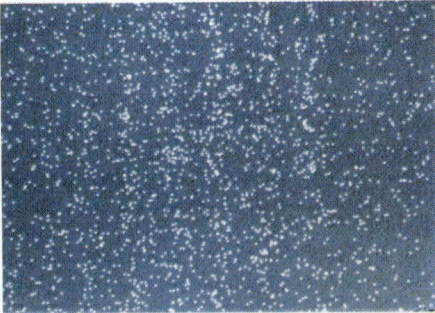

(a)

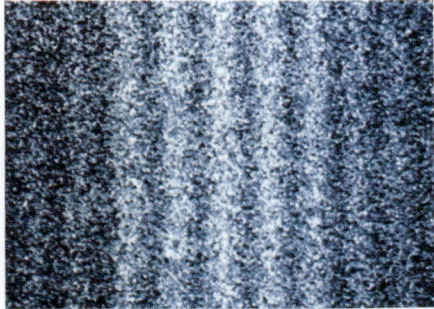

(c)

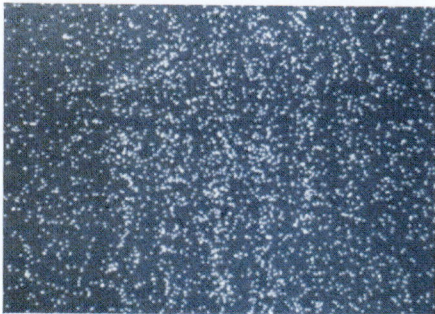

(b)

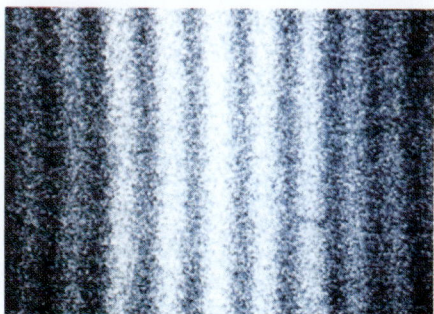

(d)

Fig. 42. Photographs demonstrating how single electrons build up the biprism interference pattern. Number of electrons detected: (a) 500, (b) 1,000, (c) 7,000, (d) 50,000.

You might think that interference fringes are produced only when two electron waves pass through both sides of the electron biprism simultaneously. If there were two electrons in the microscope at the same time, such an interference might happen. But this cannot occur, because there is no more than one electron in the microscope at one time even when the most coherent electron beam now available is used, and because an electron has never been found to be split into two or more pieces.

Let us watch what happens a little later on the monitor shown in Fig. 42. This time the whole field of view on the monitor is observed. When a large number of electrons is accumulated, something like regular fringes begin to appear in the vertical direction as shown in Fig. 42(a) and (b). You can recognize the interference fringes in Fig. 42(c). Note that the interference fringes are made up of bright spots, each of which records the detection of an electron. The last scene shows high-contrast biprism interference fringes [Fig. 42(d)].

We have reached a very strange conclusion. Interference fringes were observed even though electrons were sent through the apparatus one by one. These interference fringes are produced only when electron waves pass through both sides of the biprism at the same time.

Electrons are always detected as single particles. When accumulated, however, the interference fringes are formed. Recall that at any instant there was at most one electron in the microscope. We have reached a conclusion which is far from what our common sense tells us.

Where Did Electrons Pass Through?

We have no experimental evidence that electrons passed through either one of the two sides of the biprism. We don't know whether electrons passed through the right hand side or the left hand side of the biprism, since electrons were not caught in the act. We cannot determine the positions of electrons unless we observe them. We only infer that, because electrons are always detected as particles, they must behave as particles even when we are not looking at them. We have arrived at a point, however, where the experimental results cannot be explained by inference based on our common sense.

Let us consider what happens when one side of the biprism is closed as shown in Fig. 43(b). Electrons having passed through the open side of the biprism form a uniform intensity distribution in the observation plane. The same is true when the other side of the biprism is open.

However, interference fringes are formed when both sides of the biprism are open [Fig. 43(a)]. In this case, there is a point in the interference pattern where the electron intensity goes to zero and no electrons can arrive, as, for example, at point P in Fig. 43(a). But electrons do arrive at that point when one side of the biprism is closed [see Fig. 43(b)].

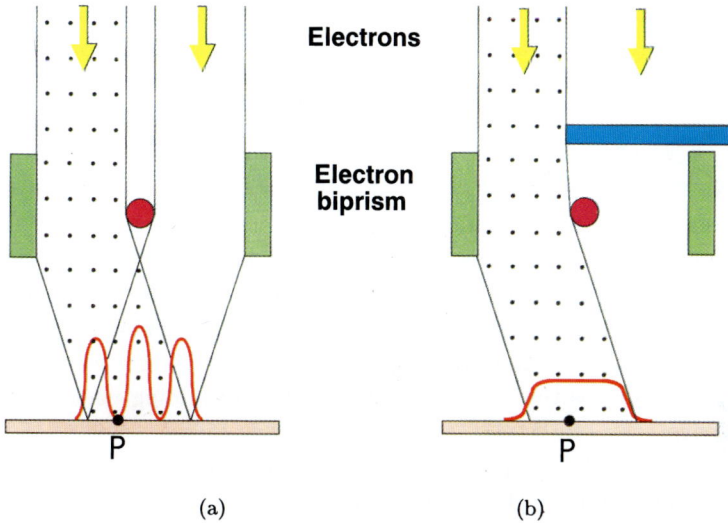

Fig. 43. Intensity distributions of electrons in a two-slit experiment.
(a) Both sides of the biprism are open.
(b) Only one side of the biprism is open.
An interference pattern is formed even when photons or electrons are sent to two slits one by one. They do not arrive at point P located at the trough of the interference pattern in the case of (a) where two slits are opened. When one of the slits is closed as in case of (b), however, electrons can arrive at point P. We can conclude that an electron did not pass through one of the two slits when both slits are open, since if the electron passed through one slit, there should have existed a probability for the electron to arrive at point P as in case of (b).

"Why are electrons forbidden from arriving at point P when both sides of the biprism are made open?" We have no choice but to conclude that electrons did not pass through only one side of the biprism when both sides were open. If some electrons did pass through the left hand side of the biprism in Fig. 43(a), then the results should have been the same irrespective of whether or not the right hand side was open or not, and consequently electrons should have arrived at point P.

Interpretation by Quantum Mechanics

"How can quantum mechanics explain this strange behavior of electrons?" — In classical mechanics, an electron is a tiny particle having a mass and an electric charge. Its behavior is described by Newton's equation of motion. When a rocket is fired from a certain place at a given velocity, we can predict when and where the rocket is flying at any given time.

The two-slit experiment, however, can never be explained by Newton's equation of motion. If electron trajectories could be traced, the interference pattern — which is produced only when two electron waves pass through on both sides of the biprism at the same time — should not be produced since each electron passes through either side of the biprism.

In quantum mechanics, electrons are considered to behave sometimes like particles and sometimes like waves. As soon as an electron is fired from the source, it begins to spread out and to propagate as a *wave* that can pass through both sides of the biprism. The electron wave is represented by a mathematical quantity, called *wavefunction* Ψ, whose behavior is determined by the Schrödinger equation. When detected, however, electrons are always detected as indivisible particles.

"What happens to the wavefunction when the electron is detected?" The wavefunction spread out in a space is considered to shrink into a spot as soon as it is detected. The wave should be concentrated at the spot. However, we cannot predict where the wavefunction shrinks. We can instead only know the probability of finding electrons at a certain point. The square of the absolute value of the wavefunction, $|\Psi|^2$, provides the probability distribution of finding where electrons are found to arrive.

In classical physics, whether Newtonian mechanics or electromagnetism, the future is uniquely determined when the initial conditions are given. In the quantum mechanics governing the microscopic world, though, the future cannot be predicted uniquely. In fact, the photographs shown in Fig. 41 cannot be explained by quantum mechanics. What quantum mechanics can predict is only the probability concerning where electrons are going to be detected. What we can say about the photographs in Fig. 41 is that although electrons look to be distributed quite randomly, the distribution *should* follow the probability distribution $|\Psi|^2$. That means the results in this quantum world are governed by chances.

What are Electron Waves?

You may not understand what electron waves are. You may think that inasmuch as electrons produce interference fringes just like water waves do, electrons must somehow oscillate periodically with time and propagate in a space.

Waves on the water surface seen in Chapter 2 are produced by the oscillation of the water surface up and down. The shape of the water surface can be observed through the transparent side of the ripple tank (see Fig. 11), and this shape can be mathematically expressed using the distance y displaced in a vertical direction from the average surface level: In the case of a plane wave, y can be expressed by a sine or cosine function. For example,

$$y(x, t = 0) = F \cos kx, \tag{7}$$

where F is the maximum displacement of the water surface, k is a constant, and x is the coordinate (distance) along the direction of the wave propagation. This equation describes the wave on the water surface at a moment $t = 0$ [see Fig. 44(a)].

"What about the wave at nonzero values of t?" If this wave is assumed to travel with a constant velocity v in the x-direction, the wave after t seconds is the same as the wave at $t = 0$ when x is replaced by $x - vt$. Therefore,

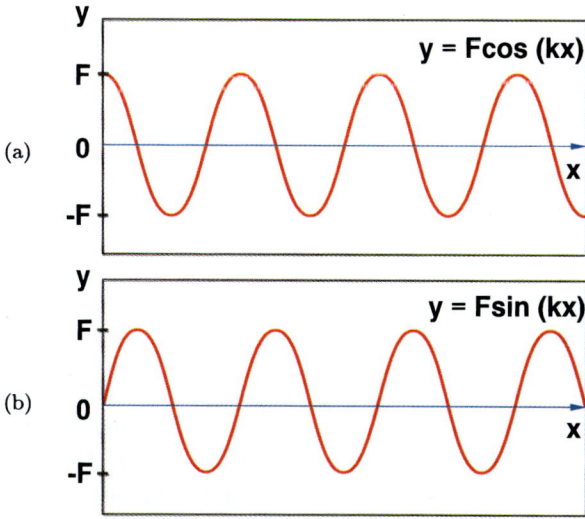

Fig. 44. A electron plane wave composed of real and imaginary waves.
    (a) Real wave
    (b) Imaginary wave
    An electron wave consists of the real wave and the imaginary wave. The two waves are phase-shifted by $\pi/2$. Since the intensity of the total wave is given by the summation of the intensities of the two waves, *i.e.* $|y|^2 = F^2 \cos^2 kx + F^2 \sin^2 kx = F^2$, which is constant, we cannot observe the wavefronts (peak positions) of the electron wave.

$$y(x, t) = F \cos(kx - \omega t). \tag{8}$$

where $\omega = kv$.

"How can the *electron wavefunction* be expressed?" Inasmuch as the electron wave is a kind of wave, we would think it must be described mathematically by an equation like (8). This is not the case, however. Electron waves are beyond our common sense and are essentially different from the water waves or string waves we already know. The electron wavefunction cannot be expressed as in Equation (8). The wavefunction should be a solution of the Schrödinger equation, but Equation (8) cannot be that solution.

Wavefunction Described by Complex Numbers

The wavefunction can be expressed not by *real numbers*, but by *complex numbers*. "What are *complex numbers*?" The physical quantities we can measure, such as the distance the water surface goes up or down, are all expressed by real numbers. Both positive and negative numbers are real numbers.

Complex numbers, on the other hand, are, so to speak, numbers in a space different from ours. A complex number is the sum of a real number and an *imaginary number*. An electron wave is expressed as the sum of a *real wave* and an *imaginary wave*.

Let us consider a plane wave of electrons. A water plane wave is expressed by $y(x) = F \cos kx$ as shown in Equation (7), which is not a solution of the Schrödinger equation. "How can we get a wavefunction for an *electron* plane wave?" It can be obtained by adding to Equation (7) an imaginary wave expressed by a cosine function phase-shifted by $-\pi/2$ [Fig. 44(b)]; *i.e.* $\cos(kx - \pi/2) = \sin kx$:

$$y(x, t = 0) = F(\cos kx + i \sin kx), \tag{9}$$

where i represents an imaginary number, $i = \sqrt{-1}$. The product of i and i is -1.

Unobservable Wavefronts

The wavefronts of a water wave can be observed, for example, as lines connecting the points where $y(x, t)$ in Equation (8) has the peak value. On the other hand, the wavefronts of an electron plane wave cannot be observed. This is because what we can observe about the electron wave is not given by the intensity of a real wave but by the sum of the intensities of a real wave and an imaginary wave, *i.e.* $|\Psi|^2$: the crests of a real wave and the crests of an imaginary wave are always displaced from each other by $\pi/2$ and the observable quantity $|\Psi|^2$ does not change with position x but takes a constant value.

Let me explain this in a little more detail. You are familiar with the absolute value of a real number: if the real number is positive, the absolute value is equal to the number. If the real number is negative, the absolute value is the positive number obtained by reversing its sign. Therefore, the square of a real number is equal to the square of the absolute value of the real number.

The situation is different for a complex number, $\Psi = a + ib$, where a and b are real numbers. Its square is also complex: $\Psi^2 = a^2 - b^2 + 2iab$, but the square of the absolute value of a complex number $|\Psi|^2$ is always positive, since it is defined as the sum of the squares of its real part and its imaginary part:

$$|\Psi|^2 = a^2 + b^2. \tag{10}$$

Since $|\Psi|^2$ always takes a positive value, $|\Psi|^2$ can be interpreted as the probability distribution of finding an electron at position x. If the value of $|\Psi|^2$ is complex or negative, it cannot be interpreted as the probability. Let us calculate $|\Psi|^2$ for an electron plane wave given by Equation (9):

$$|\Psi|^2 = F^2(\cos^2 kx + \sin^2 kx) = F^2. \tag{11}$$

This equation indicates that the quantity which we can observe, $|\Psi|^2$, is a constant independent of position x. That is, we can observe neither crests nor troughs of the wave expressed as Equation (9).

Some of you may doubt whether these wavefunctions can really represent waves, since Equation (11) indicates the fact that we cannot observe the wavefronts of an electron plane wave. Evidence for electron waves appears when two electron waves overlap.

Interference of Electron Waves

Let us see what happens when we use an electron biprism to make two electron waves overlap (Fig. 21). When an electron plane wave enters the biprism, two plane waves mutually tilted toward the center leave the biprism. Each of the two plane waves consists of its real wave and imaginary wave.

The two plane waves having passed through both sides of the biprism overlap in the observation plane. The corresponding two wavefunctions, Ψ_1 and Ψ_2 are added there. The summation of two complex numbers is given by the separate summations of their real parts and imaginary parts:

$$\Psi_1 + \Psi_2 = (a_1 + ib_1) + (a_2 + ib_2) = a_1 + a_2 + i(b_1 + b_2). \tag{12}$$

That summation is of course complex and is also a solution of the Schrödinger equation. If you add any number of solutions of the Schrödinger equation, the result is also a solution of the Schrödinger equation.

Now let us consider the intensity distribution of accumulated electrons. You can get it by calculating $|\Psi|^2$:

$$|\Psi|^2 = |\Psi_1 + \Psi_2|^2 = (a_1 + a_2)^2 + (b_1 + b_2)^2. \tag{13}$$

This distribution gives the same interference pattern as in the case of water waves, which will be explained in the following section.

As soon as electrons leave the source, they disappear in the form of wavefunctions expressed by complex numbers. When detected, however, electrons appear as particles and when they are accumulated, the intensity distribution becomes the same as that of the interference pattern of real waves such as water waves. There is thus no doubt that electrons have the wave nature.

Real Wave and Imaginary Wave

"Why does the interference pattern of electrons become the same as that of water waves?" To physically understand this question, let us consider how electron waves, after having passed through both sides of the biprism, interfere with each other in the observation plane. The right and left waves each consist of a real wave and an imaginary wave, and the real wave and the imaginary wave are always phase-shifted with each other by $\pi/2$.

First we consider the pattern produced by the interference of the two *real* waves, one having passed through the left hand side of the biprism and the other through the right hand side. The two real waves with amplitudes a_1 and a_2 in Equation (12) overlap *in phase* at the center of the observation plane to cause a vigorous oscillation there. This interference pattern is the same as that of water waves shown in Fig. 13. The intensity distribution $(a_1 + a_2)^2$ represents regular fringes with equal spacing as in Fig. 13(b).

"What about the interference pattern between the two *imaginary* waves?" Although the real wave and the imaginary wave of a wave passing through one side of the biprism are *not in phase* but phase-shifted by $\pi/2$, the two imaginary waves passing through two sides of the biprism are *in phase* at the center. Therefore, the resultant interference pattern $(b_1 + b_2)^2$ is also the same as that of the water wave as in Fig. 13(b).

In summary, the interference pattern formed by the two *real* waves is the same as that formed by the two *imaginary* waves, even though the real and imaginary waves are always phase-shifted by $\pi/2$.

Now the situation has become clear. If you recall that the intensity of a complex electron wavefunction is given by the sum of the two intensities of the real and imaginary waves [Equation (13)], it is evident that the electron interference pattern is equivalent to that of water waves. The electron wavefunction behaves as if a real wave interferes only with another real wave and an imaginary wave interferes only with another imaginary wave. There is no interference between a real wave and an imaginary wave. The resultant intensity of an electron wave is the sum of the intensities of the real wave and the imaginary wave. We have thus come to the conclusion that the interference pattern of electrons is the same as that of water waves, though electron waves differ from water waves in that the amplitude of an electron wave is a complex number.

The Distribution of Electron Arrival-times

Up to now, we have paid attention only to the spatial distribution of the electrons, such as the biprism interference pattern in Fig. 42(d). We have not been concerned with the *arrival times* of the electrons. In the two-slit experiment, however, not only the positions but also the times of electron arrival were observed and recorded on video tape. "At what intervals are electrons detected when we actually measure the times of electron arrival at a point on the detector?" Electrons appear to arrive at random intervals.

"What happens if the arrival times are measured more precisely? Are they still at random?" Electrons behave as waves. Therefore, when two electron waves accidentally overlap, some effect must occur due to the interference between the two electron waves thus producing an influence on the distribution of the arrival times.

We cannot detect such an interference effect by observing a conventional interference pattern. Instead we have to precisely measure the arrival times of electrons to see which of the three cases in Fig. 45 actually takes place. Namely, we study the correlation between successive arrival times. Figures 45(a), (b) and (c) correspond to the cases of zero, positive and negative correlation.

Such events whereby two electrons in a coherent electron beam overlap must surely happen though it is extremely rare. According to the prediction of quantum mechanics, the distribution of the arrival times should deviate slightly from the random distribution because of the overlap of two electrons. It is furthermore predicted that the deviations from the random distribution are in the opposite directions in two cases of light and electrons! It has become a more interesting story. We have

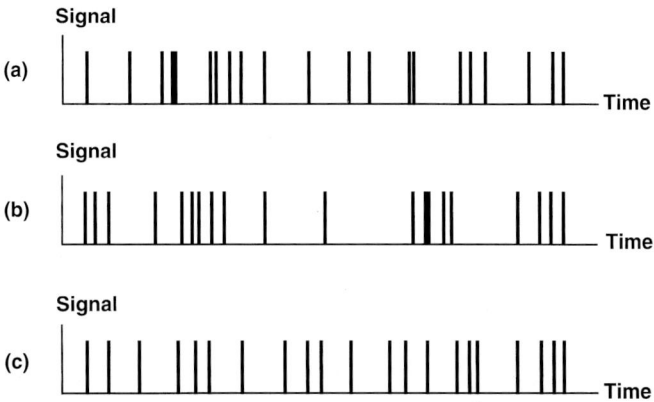

Fig. 45. Arrival times of particles (from R. Loudon, The Quantum Theory of Light, 2nd ed. (Oxford, New York, 1983)).
(a) Random arrival times (zero correlation)
(b) Bunched arrival times (positive correlation)
(c) Antibunched arrival times (negative correlation)

no way of distinguishing whether interference patterns are formed using electrons or light as far as we are looking at the interference patterns. In the interference experiments of measuring the arrival times of particles, however, we can tell the difference between two cases of electrons and light.

It was in 1956 that the correlation between the photon arrival times in a coherent beam of light was actually detected. A controversy followed whether or not such phenomena could occur. The experiment is called *the Hanbury Brown–Twiss Experiment* after the discoverers, R. Hanbury Brown and R. Q. Twiss, in Australia.

Hanbury Brown–Twiss Experiment

Hanbury Brown and Twiss investigated the presence of the correlation between photons in a coherent beam using the experimental apparatus shown in Fig. 46. They did not directly measure the arrival times of successive photons but measured the correlation between the intensities at the two separate detectors located nearby. If photons would arrive not at random but at equal intervals, then when a photon arrived at one detector another photon could not arrive at the other detector simultaneously. On the other hand, if photons would arrive in bunches, then when a photon arrived at one detector another photon would tend to arrive at the other detector.

They found a positive correlation between the intensities at the two detectors: When one intensity became intense the other intensity tended to be intense. That is, photons came actually *in bunches*. The correlation disappeared when they put the two detectors far apart from each other. This is a clear evidence that the effect is due to interference. The correlation disappeared when the distance between the

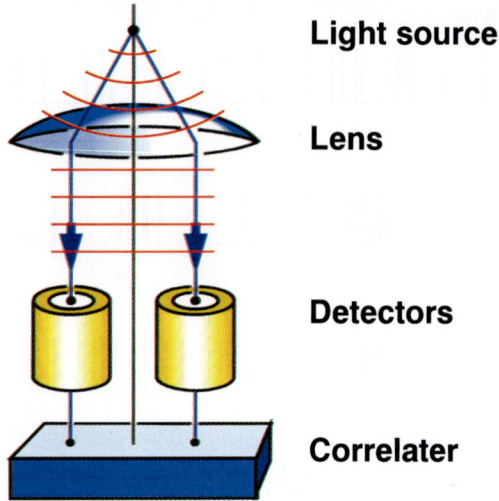

Fig. 46. Hanbury Brown–Twiss experiments.
    The intensities of coherent light at two detectors are positively correlated, *i.e.* the intensity
    of one detector tends to become high when that of the other detector is high. Photons
    arrive in bunches!

two detectors became larger than the spatial coherence length D of the light beam
incident onto the detectors given by Equation (5). They used this equation to
measure the diameter of a small star too small to be observed with telescopes.

As soon as the result asserting "Photons arrived in bunches" was published in
the scientific magazine "Nature," it created a sensation. More experiments soon
followed and two results were reported. No correlation was detected in either of
them. E. Brannen and H. J. S. Ferguson went as far as to say in their paper that *if
such a correlation did exist, it would call for a major revision of some fundamental
concepts in quantum mechanics.*

Their assertion, however, proved to be false. Hanbury Brown and Twiss, and
E. M. Purcell pointed out that under the conditions of Brannen and Ferguson's
experiment the correlation could not be detected until the experiment continued
for hundreds of years!

The intuitive explanation given by E. M. Purcell [See E. M. Purcell: "The
Question of Correlation between Photons in Coherent Light Rays" Nature **178**
(1956) 1449.] is as follows: When two photons overlap accidentally, the two waves
interfere. [Here we assume that the polarizations of photons are the same.] When
the two waves interfere *in phase*, the wave amplitude increases twofold as we have
seen in the case of string waves in Fig. 8(c). Since it is the intensity of the wave or the
number of photons that is detected, the intensity which is given by the square of the
amplitude increases fourfold. When the two waves overlap *out of phase*, however,
the amplitude and consequently the intensity vanishes as in the case of Fig. 8(d).
In this way, the intervals of the arrival times of photons must thus have a larger

fluctuation than those of a beam of random particles. P. A. M. Dirac once said "A photon interferes with itself. Interference between different photons never occurs." He was referring to the conventional observation of interference fringes. One can see the interference of two photons by observing the correlation, as in the Hanbury Brown–Twiss effect.

Now we have arrived at our problem of what happens to electrons instead of photons.

Correlation Experiment Using Electrons

As Purcell already pointed out in his paper, a negative correlation instead of a positive correlation would be produced in an electron version of the Hanbury Brown–Twiss experiment, since the overlap of two coherent electron waves is forbidden. The reason why two electrons are forbidden to overlap comes from the fact that electrons belong to Fermi particles. More than one Fermi particles are not allowed to be in the same state by the Pauli exclusion principle, while photons belong to Bose particles and more than one photon are allowed to be in the same state. "Is it possible to carry out an electron version of the Hanbury Brown–Twiss experiment?" Just after the Hanbury Brown–Twiss experiment was reported, people did not believe in such a phenomenon and similar experiments were carried out to test the result. It turned out, however, with the apparatus used in those experiments, it would have taken hundreds of years to detect the correlation. The time resolution and the counting rate should be much higher if this small effect is to be detected.

The experiments to detect the correlation of electrons are much more difficult than those in the case of photons, which themselves were very difficult as one can see the case of Brannen and Ferguson mentioned before. The difficulty of the experiments can be estimated from both the length of a *wave packet* in the propagation direction and the average distance of two successive wave packets, since we have to detect the effect of overlapping two wave packets. A wave packet is a wave corresponding to each electron. The length of an electron wave packet in a field-emission electron beam is 1 μm, 200,000 times shorter than the length, 20 cm, of a photon wave packet used in the Hanbury Brown–Twiss experiment. Interference occurs only when two electrons come as close as 1 μm, but the average distance between two electrons is 1 m at the shortest in our electron beam even when we use the brightest field-emission electron source. Since the probability that two electrons overlap is extremely small, a large number of measurements have to be accumulated to detect these very rare events.

The coherence of an electron beam, however, is a great deal better than it was 20 years ago, and therefore the probability of two electrons overlapping is a great deal higher. The problem to be left is how to detect the arrival times of successive electrons precisely. In 1986 M. P. Silverman of Trinity College in the United States stayed at our laboratory and theoretically investigated new possibilities opened by this new kind of electron interferometry.

In 1989, when an ERATO (Exploratory Research for Advanced Technology) "Electron Wavefront Project" supported by JRDC (Research Development Corporation of Japan) started, I wanted to attempt this experiment, but first we had to develop a better detector. In order to detect the correlation with the detector with a precision of 0.1 nanoseconds, we have to measure the arrival times of as many as 10 billion electrons. Fortunately, Y. Tsuchiya at Hamamatsu Photonics agreed to develop this detector. We spent two years designing the experiment, which is therefore still in progress. If the observation of two-electron interference becomes feasible, a new way of electron interferometry will be opened up.

Chapter 7

INTERFERENCE ELECTRON MICROSCOPY

The development of a coherent electron beam has made it possible to directly observe the wave nature of electrons: The electron interference patterns or electron holograms became so bright that the phase distribution of an electron wave became observable even dynamically in real time, and the precision with which the phase distribution could be measured improved from $2\pi/4$ to $2\pi/100$. These developments have opened a new way to observe and measure microscopic objects and fields by using the electron phase information.

Interference Microscopy Displaying Phase Distribution

We can observe microscopic objects with electron microscopes: an electron beam penetrates a thin sample and its image is drawn with the intensity distribution of electrons transmitted through the sample. However, the intensity is not the only information electrons receive from the sample. Electrons receive another kind of information; that is, the *phase*.

The phase information of a wave can be measured as an interference pattern. The phase information in an optical beam has frequently been used to observe or measure objects in what is called *interference microscopy*. In an optical interference micrograph where the phase distribution is displayed using interference fringes, we can directly observe the thickness distribution or the refractive index distribution of a sample.

"What can be seen by interference *electron* microscopy?" Electrons when they go into a sample, are accelerated a little bit and the wavelength becomes shorter inside it. This situation can easily be understood when you recall the state of electrons in a metal: free electrons in a metal can move freely inside but cannot go outside of it, since they are confined in a potential well just like water in the pool [see Fig. 35(a)]. Therefore, an electron going into the metal increases its velocity by an amount corresponding to the height difference between two levels of the pool and the ground in Fig. 35(a). This height difference is called the *inner potential*. The inner potential is ten to thirty volts and is specific to a material. Therefore, when an electron enters a material and increases its velocity by this inner potential, the electron wavelength becomes shorter inside it according to de Broglie's relation [see Equation (1)]. Therefore, a relative phase shift is produced between two electron waves, one passing through a material and the other passing in a vacuum. The phase shift is proportional to the distance the electron beam travels inside a material and consequently is proportional to the thickness of a material uniform in substance.

The production of an electron phase shift can also be explained by using a refractive index n. The refractive index of a material is defined as the ratio of the electron wavelength in a vacuum to that in a material. Since electrons inside the material are accelerated by its inner potential V_0, the refractive index n can be expressed using Equation (2) as follows:

$$n = \frac{\lambda_0}{\lambda} = \sqrt{\frac{V + V_0}{V}}, \tag{14}$$

where λ_0 and λ are respectively the electron wavelengths in a vacuum and in the material, and V is the accelerating voltage of incident electrons.

In the case of a 100-kV electron beam, $n - 1$ ranges from 5×10^{-5} to 1.5×10^{-4} while in the case of a light beam, $n - 1$ is much larger than that of an electron beam. For example, $n - 1$ is as large as 0.5 for glass. Therefore, a phase shift of an electron beam transmitted through a sample is much smaller than that of light.

Holographic Interference Electron Microscopy

"How can we get an interference *electron* micrograph where the phase distribution of the electron beam transmitted through a sample is displayed using interference fringes on its electron micrograph?" An interference micrograph can be obtained simply by using an electron microscope equipped with an electron biprism. But this results in an *interferogram* and not a *contour map*: the phase distribution cannot be displayed as a contour map, but only as deviations from regular interference fringes (interferogram). The highest precision in the phase measurement with this method is limited to $2\pi/4$. Using electron holography, however, we can get a contour map and in addition, the measurement precision can increase to $2\pi/100$. The reason why the precision increases is explained in the following section.

In electron holography we first form an electron hologram and then illuminate it with a laser beam to get an interference micrograph. A typical arrangement of the hologram formation is shown in Fig. 47(a). An off-axis electron hologram is formed in an electron microscope where an electron biprism is installed. An object is situated on one half of the specimen plane. An electron beam transmitted through the object passes through on one side of the biprism, and a reference beam passes on the other side. When there is no object on the specimen plane the hologram is simply a biprism interference pattern. A hologram with an object is a biprism interference pattern modulated by an object. An image of an object is often in focus in the hologram plane but this is not always necessary.

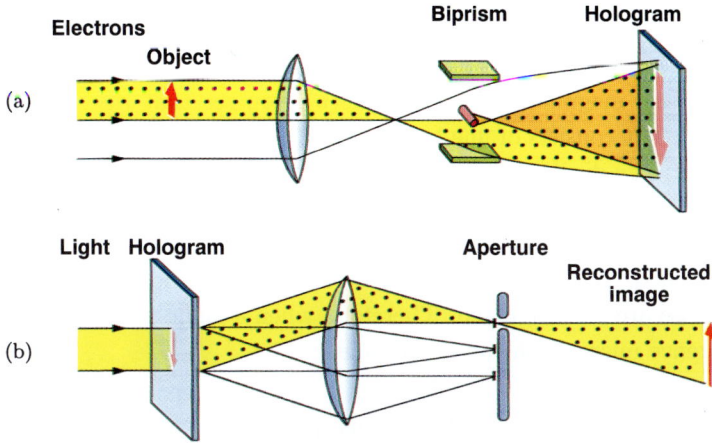

Fig. 47. Principle behind electron holography.
   (a) Hologram formation with electrons
   (b) Image formation with light
   A hologram, which is an interference pattern between an object beam and a reference beam, is formed using the electron biprism in an electron microscope. The hologram is then illuminated by a light beam to reconstruct the image in a diffracted beam.

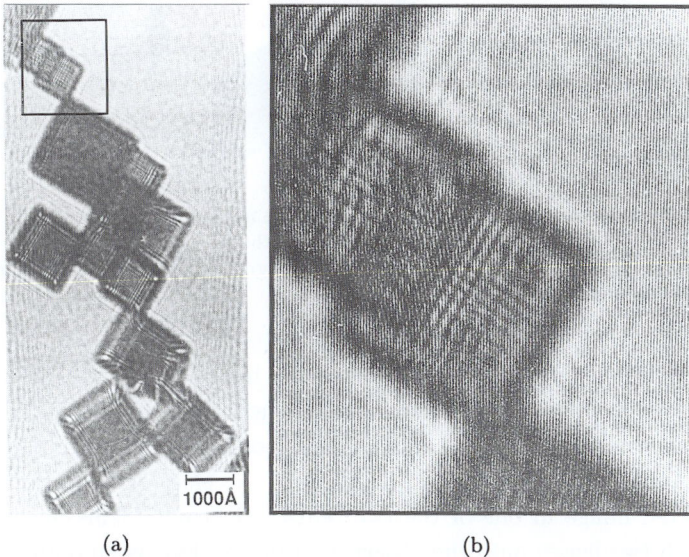

(a)   (b)

Fig. 48. An example of an electron hologram.
   (a) Electron hologram of magnesium oxide particles
   (b) Magnified hologram
   The hologram (a) simply looks like an out of focus image, but when enlarged it turns out to consist of fine interference fringes. The phase information of an electron beam is recorded as fringe displacements.

It is easy to understand what a hologram looks like when you see the example shown in Fig. 48. It resembles a simple out-of-focus image [Fig. 48(a)], but when you look at the magnified image [Fig. 48(b)] you see that the hologram is not a simple image but consists of fine interference fringes. Electron *phase* information is recorded as fringe displacements.

When a laser beam illuminates the hologram, two diffracted beams are produced [Fig. 47(b)]; in one a reconstructed image is formed and in the other a conjugate image is formed. In the optically reconstructed image, not only the intensity but also the phase are reconstructed. Therefore, when a plane wave overlaps the image, an interference micrograph can be obtained.

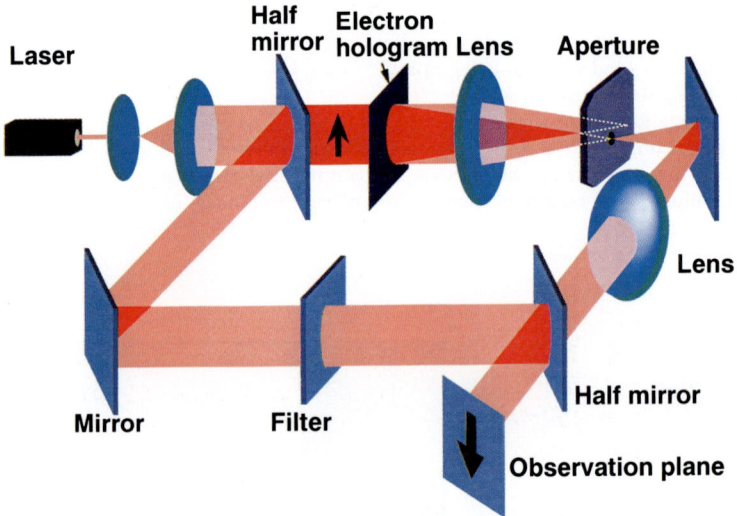

Fig. 49. An optical reconstruction system for interference microscopy.
When a light beam illuminates the hologram, the image is formed on the observation plane through two lenses. An interference micrograph can be obtained by overlapping a plane wave onto this image.

"How can we overlap a plane wave onto the image to get the interference micrograph?" An example of an optical reconstruction system for interference microscopy is shown in Fig. 49. The arrangement looks complicated, but the principle is simple. A collimated laser beam illuminates the electron hologram producing a reconstructed image in one of the two diffracted beams. This image is formed again through two lenses onto the observation plane. This is an optical image but is an exact replica of an electron image. A plane wave split by a half mirror overlaps this image to produce the interference micrograph. When the propagation directions of the object wave and the plane wave are the same, the *contour map* is obtained. When the two waves propagate in different directions, an *interferogram* is formed.

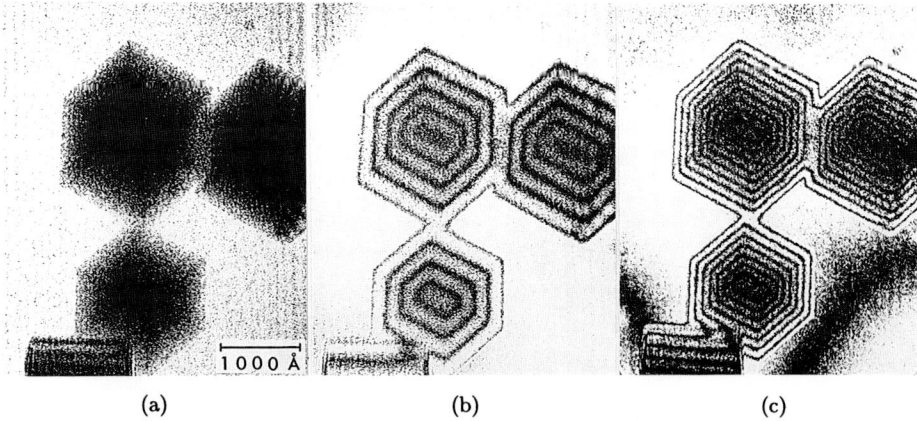

Fig. 50. Interference micrographs of magnesium oxide particles.
(a) Reconstructed image, *i.e.*, electron microscopic image.
(b) Conventional interference micrograph
(c) Twice phase-amplified interference micrograph
Only the outline can be seen in the electron microscopic image (a) where the intensity of the transmitted electron beam is observed. While in the interference micrograph (b), the phase contour lines indicate the thickness contours. The sensitivity of the phase measurement can be increased as in (c) using a technique peculiar to holography.

Examples of interference micrographs are shown in Fig. 50. The specimen here consists of magnesium-oxide smoke particles produced when magnesium is burned in air. A reconstructed image (a), which records the intensity of the transmitted electron beam, provides us only the outline of a sample, while interference micrographs (b) and (c) look three-dimensional. This is because the thickness distributions are displayed as contour fringes in the interference micrographs. One fringe spacing in micrograph (b) corresponds to a thickness change of 400 Å. You may think that this value is extremely large compared to the short electron wavelength, for example, 0.04 Å of 100-kV electrons. This low measurement sensitivity comes from a small value of $n - 1$.

We cannot measure the thickness distribution of a sample less than 400 Å thick with this method, since even a single contour fringe does not appear in its interference micrograph. But we have developed a good method for it. The interference micrograph in Fig. 50(c) gives more detailed information about the thickness distribution than the micrograph in Fig. 50(b). One fringe spacing there corresponds to 200 Å. The phase sensitivity is improved twice. "How is such a thing possible?" The principle behind this *phase amplification* will be explained below.

Measurement Sensitivity of Phase

The phase amplification is indispensable for electron interferometry. One might think that, because of the electron's extremely short wavelength, the thickness distribution of a specimen could be measured with high precision by electron interferometry. However, the fact is different. Since the refractive index is very close to 1, the measurement sensitivity is unexpectedly low. To get a phase difference of 2π, we need a thickness difference of several hundreds of angstroms. Therefore, the sensitivity about a hundred times better than that is necessary if we want to measure the thickness of an object in the atomic scale.

Fig. 51. A differential optical micrograph of spiral steps on a $NbSe_4I_{0.33}$ crystal surface (by courtesy of I. Nakada).

"How sensitive is the *optical* interference microscope?" It was reported that the smallest phase shift detectable can be as little as $1/2000$ of the wavelength of the light used. In Fig. 51 you can see the beautiful spiral atomic steps on the crystal surface observed with an optical differential interference microscope. Such a small phase shift can be detected with light. "Why not with an electron beam?" While the highest measurement precision was possible using an electron microscope equipped with an electron biprism was $2\pi/4$ at best.

In the phase-amplification technique using electron holography, a *conjugate image* is used to improve the phase sensitivity. Recall that two images are reconstructed when a light beam illuminates a hologram. In the case of in-line holography a conjugate image is the unwanted image which disturbs the reconstructed image. The phase of the conjugate image is the same as that of the reconstructed image in absolute value but is opposite in sign. In other words, the reconstructed wavefront is, with respect to the hologram, the mirror image of the conjugate wavefront. An ordinary interference micrograph displays contour lines of the phase difference between the reconstructed wavefront and a plane wave as shown in Fig. 52(a).

"What happens if a conjugate image is superimposed on the reconstructed image instead of a plane wave?" The phase difference becomes doubled, as can be seen in Fig. 52(b). Now you understand the principle behind the phase amplification, but, you might still wonder how you can overlap the twin images which are produced at mirror-symmetric positions with respect to the hologram. An example of an optical system that lets you do this is shown in Fig. 53.

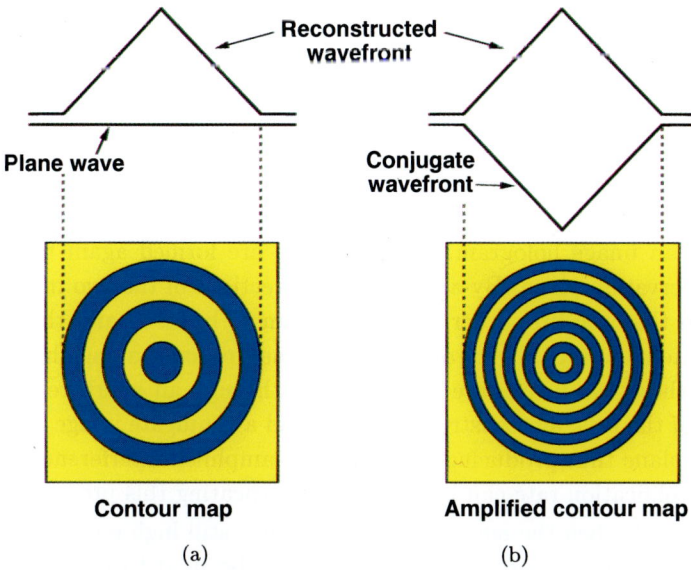

Fig. 52. The principle behind the phase-amplification technique.
(a) Conventional interference micrograph
(b) Twice phase-amplified interference micrograph
A conventional interference micrograph (a) can be obtained by overlapping a plane wave with the reconstructed image in which the phase difference between the two waves is displayed. When a conjugate image instead of a plane wave overlaps the reconstructed image, the phase difference becomes twice as large as if the phase difference is amplied twice. This is because the phase distribution of a conjugate image is reversed in sign compared with that of the reconstructed image.

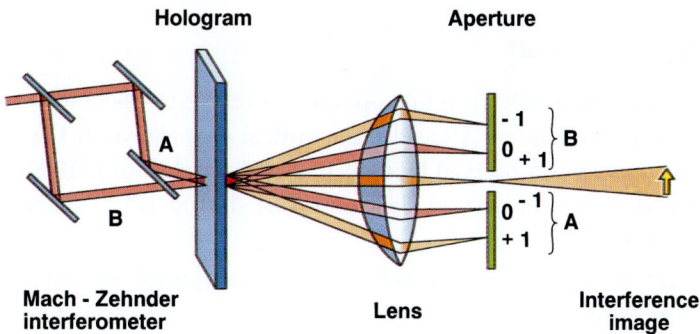

Fig. 53. An optical reconstruction system for phase-amplified interference microscopy.
Two light beams A and B are incident onto a hologram to form two sets of a reconstructed image and its conjugate through an image-forming lens. A conventional interference micrograph is obtained when the incident angles of the two beams are adjusted so that only the transmitted beam of beam A and the reconstructed image of beam B, or vice versa, may overlap by passing through the aperture located at the back focal plane of the lens. A twice phase-amplified micrograph can be obtained when only the reconstructed image of beam A and the conjugate image of beam B are selected by the aperture to overlap. The latter case is illustrated in the figure.

A laser beam enters an interferometer, called a *Mach-Zehnder interferometer*, that consists of two mirrors and two half mirrors and that produces two coherent beams whose directions can be adjusted by tilting the mirrors and half mirrors. Two light beams, A and B, illuminate an image hologram which is formed in such a way that an image of an object is in the hologram plane. Each of the beams A and B produces its own set of reconstructed and conjugate images in two diffracted beams. Therefore, four images are formed in total *in the hologram plane* since the hologram is an image hologram. These images are formed again through a lens onto the observation plane. By adjusting the directions of the two incident beams, A and B, so that the first-order diffracted beam of beam A and the minus first-order diffracted beam of the B, or vice versa, may proceed along the optical axis. When only these two beams are made to pass through an aperture placed at the focal plane of the lens, a reconstructed image and a conjugate image overlap on the observation plane thus producing a twice phase-amplified interference micrograph. A higher amplification rate can be obtained by repeating this process.

"What about when the amplification rate is not still high enough by using this method?" — Higher-order diffracted beams can be used to achieve even higher phase-amplification, since an image can be formed in the nth order diffracted beam whose phase distribution is amplified n-times. The problem is then how to produce higher-order diffracted beams: the intensity distribution of a two-beam interference pattern forms a sine wave. If this interference pattern is recorded on film, only the first order diffracted beams are produced by illuminating a light beam onto this film. Nevertheless, in actuality, higher-order diffracted beams are produced due to the deviation from the sine wave due to the nonlinearity of recording media. Therefore, it is possible to obtain strong higher-order diffracted beams if a nonlinear and high-contrast film is used as a recording medium.

Atomic Steps Observed!

"Can we use interference electron microscopy to observe a specimen on the atomic scale?" We should be able to. Consider the optical case shown in Fig. 51, where a spiral atomic step can be observed by detecting a phase shift of 1/1000 of the wavelength. There is no reason that an electron phase shift of 1/100 of the wavelength cannot be detected.

We can get an interference micrograph phase-amplified as much as we wish using the phase-amplification technique. Many fringes may simply be inserted or interpolated between the two original fringes. However, it is a different problem whether these fringes are drawn with a high precision; we had to confirm it by using a specimen that provides a definite phase difference. Since we could find no suitable samples that provided an intermediate phase difference, we had no choice but to try detecting atomic steps directly.

The results are shown in Fig. 54. Looking at the conventional interference micrograph of a molybdenite thin film in Fig. 54(a), you might think that what you

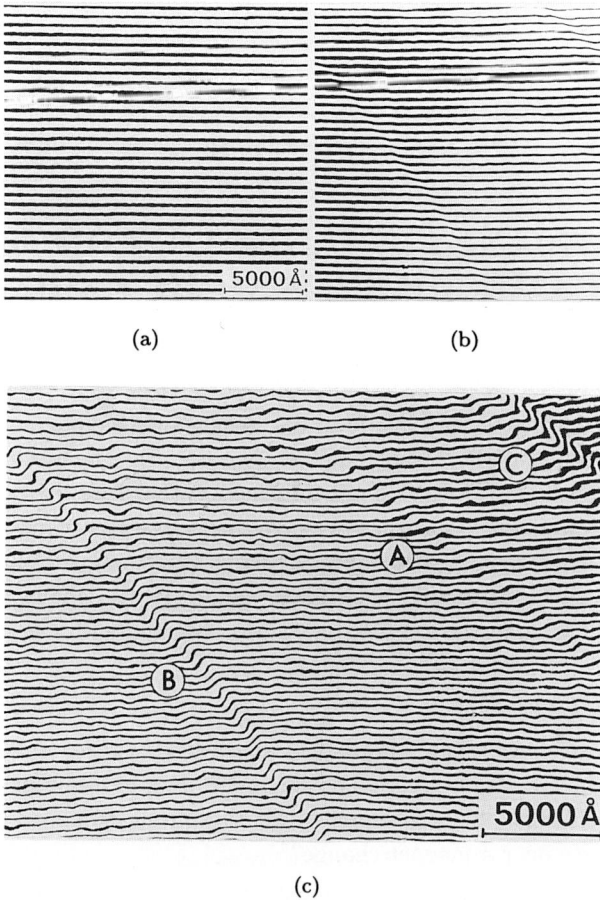

Fig. 54. An interference micrograph of MoS_2 thin film.
    (a) Conventional interferogram
    (b) 4-times phase-amplified interferogram
    (c) 24-times phase-amplified interferogram
    The phase shift of an electron beam is displayed as deviations from regular fringes in these interferograms. No deviations can be seen in conventional interferogram (a), but phase shifts due to atomic steps can be observed in phase-amplified interferograms (b) and (c). Even a monoatomic step becomes observable at A in (c).

see is simply a biprism interference pattern, since the fringes are straight. But when you look at the 4-time phase-amplified interferogram (b), you can recognize some steps. The steps are clear in the 24-time phase-amplified interferogram (c). These steps seen as displacements of regular interference fringes represent atomic steps on the film surfaces. The step at A in Fig. 54(c) indicates a monoatomic step with the height difference of only 6 Å, which corresponds to a phase shift of $2\pi/50$. Steps at B and C in Fig. 54(c) respectively correspond to three and five atomic steps.

    We could get evidence from this experiment that a phase shift as small as $2\pi/50$ could be detected!

Chapter 8

MAGNETIC LINES OF FORCE
IN THE MICROSCOPIC WORLD

At the beginning of this book, I introduced a picture showing magnetic lines of force leaking outside from tiny horseshoe magnets (see Fig. 5). The picture was made by using electron waves, but you were not informed of the reason why magnetic lines of force could be seen with electron waves. We have now arrived at the point where we can explain the reason. This chapter explains what magnetic lines of force are and why they can be observed by electron waves. It also shows you pictures of magnetic lines of force in the microscopic world observed with electron waves.

Maxwell's Image of Magnetic Lines

James Clerk Maxwell was deeply impressed with Faraday's idea of lines of force and formulated the behaviors of electric and magnetic lines of force mathematically as the Maxwell equations. He also searched for the physical image of lines of force. "How does a space near a magnet change?"

I introduce Maxwell's concept of magnetic lines of force here, because similar concepts will appear in Chapters 9 and 10 in relation to *vector potentials*. According to his thoughts, a magnetic line is produced by some fluid circulating around it. A bundle of magnetic lines is a magnetic tube. Since the fluid rotates inside a magnetic tube, its diameter tends to expand because of the centrifugal force, while its length tends to shrink.

Some of you may ask "What happens to the fluid at the boundary surface between two adjacent magnetic tubes where two streams of fluid flow in the opposite directions?" Maxwell was puzzled by this question, and gave a mechanical explanation using an *idler gear*: to make two adjacent cogwheels rotate in the same rotation direction, an idler gear is often inserted between them. When the state of magnetic tubes changes with time, then the idler gear has to move around — which movement, Maxwell thought, became an electric current. His thoughts had become unlikely since the existence of *ether* (media existent even in a vacuum through which light travels) had been ruled out by Michelson and Morley's famous experiments measuring the velocity of light.

In the present theory of electromagnetism, the concept of magnetic lines of force is not always widely used. It is magnetic fields that are used instead. Magnetic lines sometimes lead to inconsistencies if magnetic lines are to be physical objects in three-dimensional space. Such an example is discussed in *The Electromagnetic Field by A. Shadowitz* (pp. 193–194, McGraw Hill, 1975). Circular magnetic lines are produced around a straight current-carrying wire as shown in Fig. 55(a). Let's consider what happens to magnetic lines of force when a weak magnetic field parallel to the wire is added. As soon as even an extremely weak magnetic field is applied, a circular magnetic line is disconnected and becomes a long spiral [Fig. 55(b)]. Consequently, the density of magnetic lines of force suddenly becomes high. It should be noted that this infinitely long spiral shown in Fig. 55(b) is loosened from one ring in Fig. 55(a) and that such innumerable spirals appear along the wire. Thus, we have arrived at a strange result: the density of magnetic lines of force changes drastically even though the physical situation does not vary greatly.

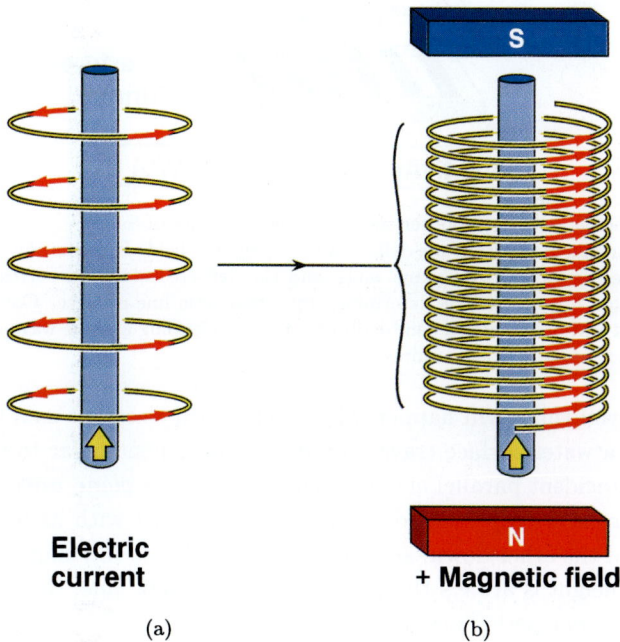

Electric current

(a)

+ Magnetic field

(b)

Fig. 55. Magnetic lines of force around a current-carrying wire.
(a) Magnetic lines of force circulating around the wire
(b) Spiral magnetic lines of force produced when a slight magnetic field is applied parallel to the wire

Magnetic lines of force in three dimensions sometimes accompany inconsistencies. When magnetic lines of force are projected in one direction, however, no inconsistencies are produced. The same is true of vector potentials. Magnetic lines of force become observable when we project them by using an electron wave.

Interference Microscopy Displaying Magnetic Lines of Force

In 1980, just after we developed a *coherent* field-emission electron beam, electron holography took a new turn. Magnetic lines of force were demonstrated to be directly observable through electron holography.

Let us consider the reason why magnetic lines of force can be observed using an electron wave (Fig. 56). Incident parallel electrons onto a uniform magnetic field are subject to a Lorentz force perpendicular to the magnetic field and are deflected.

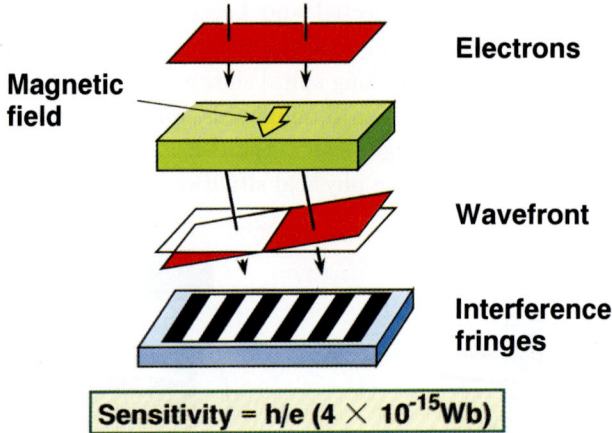

Electrons

Magnetic field

Wavefront

Interference fringes

Sensitivity = h/e (4 × 10⁻¹⁵Wb)

Fig. 56. The principle behind the observation of magnetic lines of force.
Parallel incident electrons are deflected by a magnetic field. In terms of waves, the incident electrons correspond to a plane wave, and the deflected electrons correspond to a plane wave tilted around an axis determined by a magnetic line of force. Contour lines of the deflected wavefront, which can be obtained by overlapping a plane wave (white), indicate projected magnetic lines of force.

"What do you think will happen when you regard electrons as a wave?" Recall that waves on a water surface travel in a direction perpendicular to the wavefronts (see Fig. 12). Incident parallel electrons correspond to a *plane wave*. Likewise, the deflected electrons are represented by another plane wave with its wavefront tilted, one side up and the other down, around an axis parallel to the magnetic field. Since the wavefront height is always the same along a magnetic line, the contours of this wavefront follow magnetic lines of force.

"How can we observe the wavefront contours?" Suppose the tilted plane wave could be superimposed with another plane wave with a horizontal wavefront (a white one in Fig. 56) which we use as a "reference wave." At the intersection of the two wavefronts, the phase difference vanishes and the two waves interfere constructively, producing a bright fringe on the observation plane. The phase difference increases in proportion to the distance from the intersection. When the phase difference reaches 2π, the interference becomes constructive and forms a bright fringe again. In this way, an array of bright fringes is produced. These fringes are the wavefront contours and follow the magnetic lines of force. This cannot be done in electron microscopes, but can be done by holography using the optical technique shown in Fig. 49.

Thus, we have reached a very clear conclusion that the contour fringes in the interference micrograph represent projected magnetic lines of force. In addition, quantitative information can be derived by solving the Schrödinger equation: consider two electron beams starting from a single source, passing through two points A and B in the magnetic object, and then recombining at another point by the action of the biprism (see Fig. 57). The Schrödinger equation tells us that the phase difference between these two beams is proportional to the magnetic flux enclosed by the two beams even when the path difference between the two beams vanishes.

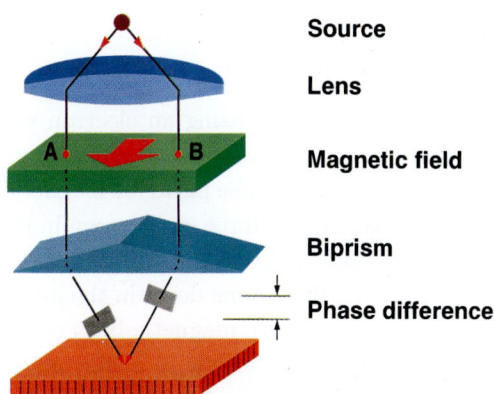

Source

Lens

Magnetic field

Biprism

Phase difference

Fig. 57. The production of an electron phase shift due to magnetic field.

If the two beams enclose no magnetic flux — that is, if the two points A and B are along a magnetic line of force — then no phase difference is produced. Consequently, *phase contours are along magnetic lines of force.* If the two beams enclose a magnetic flux of h/e, the phase difference becomes 2π. *Contours appear at every h/e flux unit.*

Electron holography can thus be used to make measurements, not only to see magnetic lines of force in a magnetic sample, but even when the sample is minute.

Magnetic Lines of Force Inside a Magnet

From now on, I am going to introduce examples of magnetic lines of force in the microscopic world observed with an electron wave. As Faraday conjectured, magnetic lines existed not only outside a magnet but also inside it. When you actually see the magnetic lines inside a magnet, their behaviors are quite interesting.

Let us first take as an example the iron filings Faraday used to observe magnetic lines of force. Iron filings are like grains of sand and do not attract one another. They become magnets and stick to one another only when a magnetic field is applied.

"Why do iron filings become magnets when a magnetic field is applied to them?" An iron filing is composed of tiny magnet regions with different orientations. A small magnet region is called a *magnetic domain*, and the boundary between domains is called a *domain wall*. Because these tiny domains are usually oriented randomly, an iron filing as a whole does not behave like a magnet. No magnetic fields leak outside an iron filing. Therefore, even when iron filings are brought close to one another, nothing happens. When iron filings are brought into a magnetic field, however, they instantly become magnets and begin to attract one another. The way in which an iron filing becomes a magnet is quite interesting: a certain magnetic domain inside the iron filing, the one whose orientation is closest to that of the external magnetic field, expands. It expands by the smooth movement of the surrounding domain walls, and the filing becomes a single-domain magnet. This is not an imaginative explanation but a true story.

Look at a Domain Wall! — Néel Wall

Let us look at an actual domain wall by using an electron wave. The photograph shown in Fig. 58 is an interference micrograph of a thin film made of an iron-nickel alloy, *permalloy*, which is often used to make parts for magnetic devices. The film is 300 Å thick, and many magnetic lines of force can be seen in the picture. They look like streams of water. In the upper half of the picture magnetic lines are flowing from right to left. When magnetic lines come down in the middle, they change their directions to the right. The region where magnetic lines change their directions is the domain wall. Magnetic lines of force flow in a film plane. Some of you may wonder if the actual magnetic lines of force should be three-dimensional. Look at a scale bar indicating one micron in the figure. Because the film is only 300 Å (three hundredth's of a micron) thick, the lines of force flowing within it can reasonably be considered to be confined to a plane. Such a domain wall, where magnetic lines rotate in a film plane, is called the *Néel wall*.

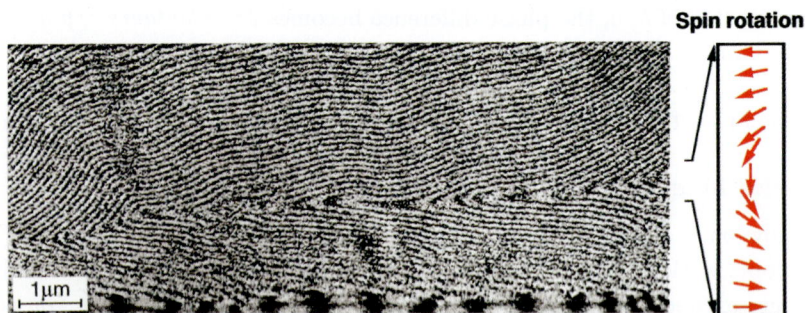

Fig. 58. An interference micrograph of permalloy film — Néel wall.
In-plane magnetic lines of force of a permalloy film as thin as 300 Å can be directly observed as contour fringes of the interference micrograph. A constant magnetic flux of h/e is flowing between two adjacent fringes. Magnetic lines of force flowing to the left in the upper region change their directions in the middle and flow in the right direction in the lower region.

In Fig. 5, we saw magnetic lines of force extending only from the north pole of a magnet to the south pole. Some of you may have thought that some kind of sources producing these magnetic lines of force must exist inside the magnet. But the magnetic lines observed inside the magnet are similar to those outside the magnet. We cannot see from these pictures how magnetic lines are produced. "Are magnetic lines of force really produced, as Maxwell imagined, by some invisible fluid circulating around them?"

How are Magnetic Lines Produced?

Magnetic lines of force must be produced from inside a magnet. "Then how are magnetic lines produced?" — They are produced from tiny magnets, *i.e.* electrons, oriented in one direction.

"How can an electron be a magnet?" An electron is a charged particle, and rotates around its own axis and has a *spin* angular momentum as I explained in Chapter 6. Because an electron is a charged particle, a rotating electron can be considered to become a tiny magnet.

A piece of permalloy contains many atoms, and each atom has many electrons. Most electrons in an atom are arranged in pairs, spin up and down, so their magnetic fields cancel one another. Atoms in materials called *ferromagnets* contain lonely electrons whose magnetic fields are not canceled. The spin directions of such electrons are aligned in a domain producing magnetic lines of force. Here, it should be noticed that although the directions of magnetic lines and spins were explained to be the same, the fact is that the two directions are opposite since an electron has a negative charge. I avoid complication at the sacrifice of the exact description.

If we accept that magnetic lines of force are produced by rotating electrons, Maxwell's conjecture cannot be said to be necessarily false. You may ask, "Why are spins aligned?" When you think of spins as bar magnets you might think the most stable arrangement of two magnets has them oriented in opposite directions. The situation is different. A quantum-mechanical force much stronger than a magnetic force, called *exchange force*, acts between two neighboring electrons and aligns electron spins. The details of this force cannot be explained any further here. If you want to know more about it you will have to study quantum mechanics.

"How do the directions of spins change at the domain wall? Do they change their directions suddenly?" If you look closely at the picture in Fig. 58 magnetic lines of force seem to change their directions gradually. This behavior of gradual change in spin direction is closely related to the smooth movement of a domain wall.

Movement of Domain Walls

Neighboring spins tend to align parallel with each other because of the exchange force. Consequently, spins cannot change their directions suddenly but only gradually so that the change in the spin direction may be shared with many spins equally.

"What happens to the domain wall in Fig. 58 when a magnetic field is applied to the thin film in the left direction?" Interference micrographs have now become observable in real-time with a recently developed technique, so we can see directly what happens. The domain wall moves downwards smoothly so that the upper domain may expand. Unless there are material defects in the film, the domain wall moves smoothly.

There is a reason why a domain wall moves so smoothly. Look at the illustration at the righthand side of the picture in Fig. 58, which shows how spins gradually rotate across the domain wall. When a magnetic field is applied in the left direction, each spin changes its direction such that its orientation becomes closer to that of the applied magnetic field. In other words, every spin rotates a little bit in the clockwise direction. As a result of such spin rotations, you will notice that the domain wall has moved downwards. The principle is the same as that behind the domino effect; *i.e.* cascading one upon the other.

When a magnetic field is applied to iron filings, a certain magnetic domain — the one whose orientation is closest to that of the applied magnetic field — expands by smoothly moving the surrounding domain walls to form a single magnet. If there is a sudden change of spins — due, for example, to impurities — there is no domino effect and the domain wall is pinned down there.

Cross-tie Wall

When the thickness of a permalloy film increases to 500 Å or more, domain walls become totally different from the one shown in Fig. 58. This is because the magnetic lines of force which were confined to a plane are now freed. This new kind of domain wall was discovered by B. J. Goodenough and his colleagues in 1958. When they sprinkled iron filings over the surface of a permalloy thin-film, they observed such domain walls as something like cross-ties intersecting with a straight line, as can be seen in the right-hand side in Fig. 59. From only this Bitter pattern, they came up with the domain-wall model shown in Fig. 59. In the illustration, a domain-wall part to which magnetic powder is attached appears as cross-ties. At first glance, the magnetization configuration may look complicated, but in principle it is actually very simple and convincing.

Let us find out why magnetic lines of force inside the wall are so complicated as can be seen in Fig. 59. First, take out only the central portion of the Néel wall. The magnetic lines of force at the domain wall in Fig. 58 are in the downward direction. This domain wall can be compared to many bar magnets directed downward and placed side by side.

"What will happen when the film is thicker?" The magnetic forces of the bar magnets become stronger, neighboring magnets will repel each other and will no longer be able to maintain their orientations. You will notice that there is a more stable arrangement of the magnets. If every other magnet is reversed, the adjacent N and S poles attract each other and the arrangement becomes more stable. This

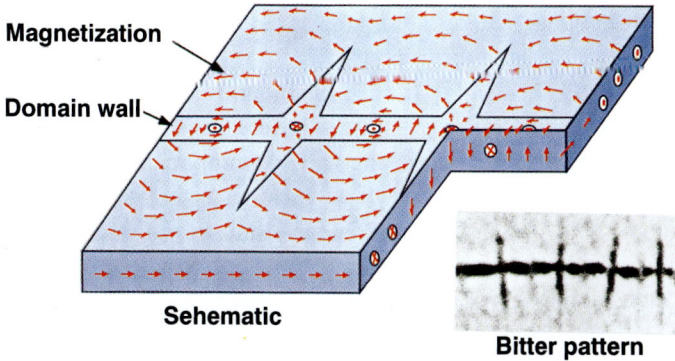

Magnetization

Domain wall

Sehematic

Bitter pattern

Fig. 59. The predicted model of cross-tie wall.

is the principle behind the cross-tie wall, and it explains why the domain wall has such a structure as that shown in Fig. 59.

Look at a Cross-tie Wall!

When we look at the interference micrograph of a cross-tie wall (Fig. 60), we can see magnetic lines of force which are exactly like the ones shown in Fig. 59. In upper and lower regions far away from the domain wall, magnetic lines of force are nearly straight, like they are in the Néel wall. The closer to the wall, however, the sharper the bend of the magnetic lines of force. Finally, they form circles and begin to rotate.

1 μm

Fig. 60. An interference micrograph of permalloy film — cross-tie wall.
   A magnetic domain wall of a permalloy thin film changes from a Néel wall to a cross-tie wall when the film thickness increases from 300 Å (Fig. 58) to 500 Å. The observed magnetic lines of force are just like those predicted (Fig. 59).

We have interpreted the interference fringes as magnetic lines of force. The fringes, however, represent the wavefront contours of an electron wave transmitted through a ferromagnetic thin film when an electron plane wave is incident onto the film. Here let us consider what the electron wavefront in Fig. 60 is like. If you are familiar with the maps of mountains, you will soon understand the shape of the wavefront from the contour map. Several mountains stand in a line. A common ridge of the mountains corresponds to the central axis of the domain wall. Unlike an ordinary mountain range where mountains stand on a plane, this mountain range stands on the top of a triangular roof.

The wavefronts are locally flat at both the mountain peaks and the saddle points between two peaks. These are the places where an incident plane wave passes through the film unaffected. It means that a component of the magnetic field parallel to the propagation direction of an incident electron wave vanishes. Since the magnitude of spin inside a magnet is constant, we can conclude that at these places the spin stands up perpendicular to the film plane. You will notice this in the Goodenough model shown in Fig. 59.

Magnetic Tape

Let me show you an example nearer at hand. It's magnetic tapes. Magnetic recording is a means of information storage used not only in video and audio tapes, but also in computer memories and in the floppy disks of personal computers. In this method, binary ("yes" or "no") information is stored as the orientation of tiny magnets. In Fig. 61(a), for example, magnets pointing to the right represent "yes," and magnets pointing to the left represent "no." We want to record more information (*i.e.* a greater number of yes's and no's) on a smaller area, and the density of the magnetic recording increases year by year in order to cope with the expanding amount of information. To find out how densely magnetic recording

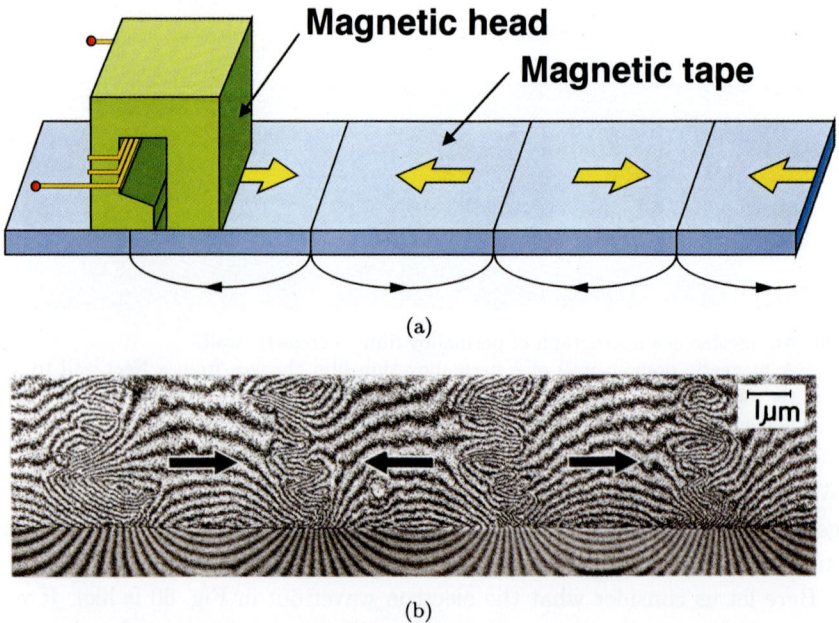

(a)

(b)

Fig. 61. Observation of magnetic lines of force recorded on magnetic tape.
(a) Method of magnetic recording
(b) Interference micrograph of recorded magnetic tape

can be made, we need to observe the recorded magnetization in detail. The recording density has recently become so high that an optical microscope cannot cope
with it. We therefore used interference *electron* microscopy to determine which are
the factors limiting high-density recording and to find out how high the recording
density can actually become.

Magnetic lines of force are recorded on a tape by a moving magnetic head whose
orientation of magnetic poles, north and south, changes [see Fig. 61(a)]. An interference micrograph of a recorded tape is shown in Fig. 61(b). The directions of
recorded magnetic lines are indicated by the arrows. When this tape is observed
using electron microscopy, nothing except the film structure can be seen. In this
interference micrograph, though, we can see many contour fringes representing magnetic lines of force. Magnetic flux flows in the directions indicated by the arrows.
They can be seen in detail even at the boundary regions where magnetic lines change
their directions. There the two opposite streams of magnetic lines collide head-on
and produce vortices just like those in flowing water. These magnetic lines meander
toward the tape edge, eventually flowing out into the air.

The problem is how densely we can pack these magnets in a tape. Obviously,
the distance between the neighboring magnets cannot be smaller than the width of
the vortex region. If we could somehow reduce the vortex size, we should be able
to reach a higher recording density.

"Can we make the vortex region narrower by changing conditions?" We can
do so by changing the tape material, the gap of the magnetic head, or the spacing
between the head and the tape. The result is shown in Fig. 62. We can see how
densely these magnets are packed now. These changes were in fact substantial. The
minimum distance between neighboring magnets shown in Fig. 62 is only 0.1 μm,
less than one tenth of the distance between those shown in Fig. 61.

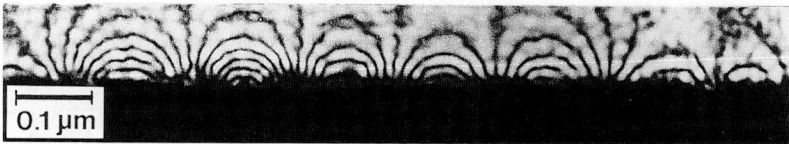

Fig. 62. An interference micrograph of recorded magnetic tape with high density (phase-amplification: ×16).

Magnetic Lines Inside Fine Particles

"What happens to iron filings when they become extremely small?" To infer what
happens, we have to know the general properties of magnetic lines of force and of
spins. As Faraday recognized, magnetic lines indicate a stream of magnetic flux
like that of water having neither sources nor sinks. A magnetic line forms a closed
loop and tends to become as short as possible. The shorter the line, the lower its
energy. The other factor influencing the behavior of magnetic lines inside magnets

is the exchange force between spins. Two neighboring spins tend to be oriented in the same direction.

An iron filing, when it's fairly large, consists of many magnetic domains. Since magnetic lines tend to shrink, the directions of magnetic lines in the domains are determined such that magnetic lines are closed inside the filing and do not leak outside. An example is shown in Fig. 63(a), where magnetic lines of force form two kinds of closed loops. In the case of a smaller iron filing, maybe it is better to call it a fine particle, only one kind of loops will be produced as shown in Fig. 63(b). This is because spins cannot change their directions suddenly, because of the exchange force.

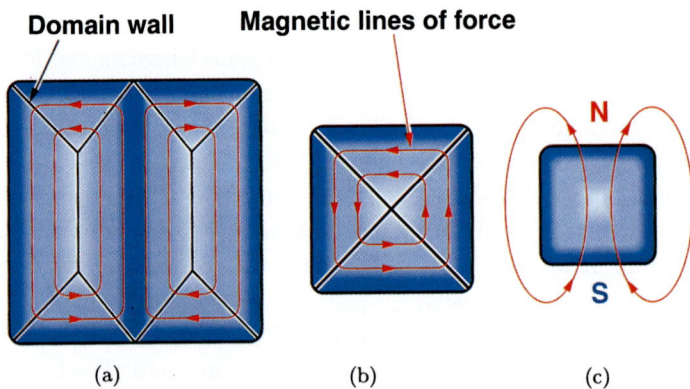

Fig. 63. Magnetic domain structures of ferromagnetic particles.
(a) Large particle having two closed domains
(b) Medium-size particle having a closed domain
(c) Singly-magnetized small particle

An observation result similar to the case in Fig. 63(b) is shown in Fig. 64, where the specimen is a fine particle of cobalt. It is plate-shaped and its thickness is uniform. In electron micrograph (a) only the outline of the sample can be seen, while in interference micrograph (b) contour fringes forming closed loops can be seen inside the particle. These fringes indicate magnetic lines of force. A minute constant magnetic flux of $h/2e$, or 2×10^{-15} Wb, is flowing between two adjacent fringes, since this interference micrograph is phase-amplified two times. As Faraday predicted, magnetic lines of force have neither beginnings nor ends and form closed loops. Magnetic lines rotate inside the particle, and do not go outside.

Perceptive people may think that the rotation direction of magnetic lines of force cannot be decided from the interference micrograph (b). It is true. The rotation direction is determined not from the contour map but from the interferogram (c), which was obtained by tilting a reference beam in the optical reconstruction stage of holography (see Fig. 49). This interferogram provides new information, information that cannot be obtained from the contour map of the electron wavefront. Specifically, the wavefront is advanced in the center of the particle, like a mountain. Since

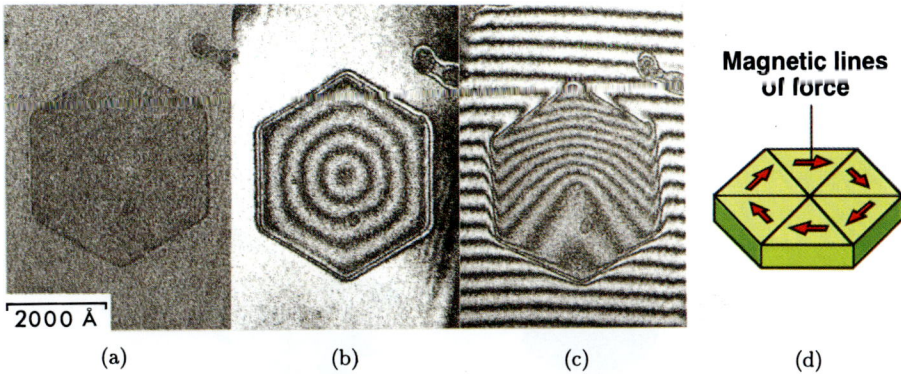

2000 Å

(a)   (b)   (c)   (d)

Magnetic lines
of force

Fig. 64. An hexagonal Co particle.
(a) Electron micrograph
(b) Interference micrograph (phase amplification: ×2)
(c) Interferogram (phase amplification: ×2)
(d) Schematic of domain structure

the wavefront is advanced or retarded depending on whether the rotation direction of magnetic lines is clockwise or counterclockwise, the direction of magnetic lines of force in this particle is decided as being clockwise from interferogram (c). If this particle is turned over, the rotation direction is reversed and the wavefront becomes retarded in the center.

"What happens when a particle is even smaller?" The particle will have a single domain as in Fig. 63(c). Magnetic lines of force have no choice but to leak outside from the particle, since the exchange force prevents the spins from changing their directions suddenly. If this particle is somehow symmetric, you may wonder why the symmetry is broken such that the N pole is produced at the upper surface of the particle in Fig. 63(c). We have four equivalent configurations. This is an example of the "spontaneous symmetry breaking," by which a specific magnetization is selected. When we have a spherical particle we have an infinite number of equivalent configurations. Other examples of spontaneous symmetry breaking will appear in the next chapter.

An example of a particle having strong magnetic anisotropy is shown in Fig. 65. The specimen is a barium-ferrite particle. Although the particle is fairly large compared with the particle in Fig. 64, it has a single domain. You can see that magnetic lines from the upper north pole of the particle sink into the lower south pole. It is interesting to see what happens when you tilt the specimen. When we look at the particle facing the surface of the north pole, we cannot see any magnetic lines of force since we are looking at projected magnetic lines of force. The direction of magnetization is determined to coincide with an easy axis of magnetization in the material.

When the particle size becomes even smaller, less than a few tens of angstroms, the particle no longer acts as a magnet. The effect of surfaces becomes enormous in such a fine particle, and consequently, the effect of thermal fluctuations of spins at the surface, becomes so large that magnetic lines of force in the particle vanish. This is explained by the theory of R. Kubo.

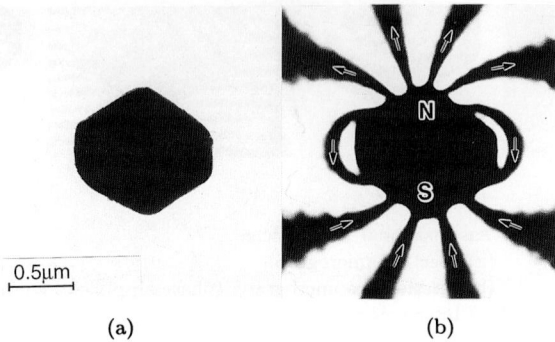

(a)   (b)

Fig. 65. A barium-ferrite particle.
(a) Electron micrograph
(b) Interference micrograph
It can be seen from micrograph (b) showing magnetic lines of force leaking from the particle that this particle has a single magnetic domain due to the strong magnetic anisotropy.

Chapter 9

THE AHARONOV–BOHM EFFECT

Around 1980, when we had developed a coherent electron beam and were absorbed in the experimental observation of magnetic lines of force in the microscopic world, there arose a controversy in the fundamentals of quantum mechanics. The problem stimulating heated discussions was a phenomenon called the *Aharonov–Bohm effect* or, simply the *AB effect*. Since this AB effect is closely related to the fundamental principle behind the method we developed for observing magnetic lines of force, we tried to confirm it experimentally. The dispute was so hot that we had to spend six years on this series of experiments. We brought this long controversy to a conclusion, at least concerning the existence of the AB effect, and demonstrated the physical reality, or to express it more carefully, the physical significance of *vector potentials*.

Prediction of the AB Effect

The AB effect is an effect of a magnetic field upon an electron which passes away from the field, and was pointed out theoretically in 1959 by Y. Aharonov and D. Bohm, then at Bristol University in Great Britain. Some of their results, that is, the magnetic AB effect had been noticed 10 years earlier by W. Ehrenberg and R. E. Siday.

If you have studied quantum mechanics, you may know the name of D. Bohm (Fig. 66) from his textbook or from the Bohm–Pines theory. After he had moved to Bristol University from the United States, his student Aharonov (Fig. 67) and he published a paper entitled, *Significance of Potentials in Quantum Mechanics* in an American journal called *Physical Review*, and the phenomena described in that paper were later called the AB effect. They devised two kinds of the AB effect, one electric and the other magnetic, but now only the magnetic one is often referred to as the AB effect.

The magnetic AB effect is illustrated in Fig. 68. When a current is applied to a coil, the coil becomes an electromagnet. North and South poles are produced at the ends of the coil. But when we think of an infinitely long coil,

Fig. 66. David Bohm (1917–1992).

Fig. 67. Yakir Aharonov (1932–).

the poles are infinitely far apart and no magnetic fields exist outside the coil. Two electron beams starting from a point source and passing on opposite sides of the coil are made to overlap with an electron biprism so that they form an interference pattern. We can see a phase difference, or a displacement of the two wavefronts between the two beams passing on the different sides of the coil.

You will think that since both beams pass through only a space free of electromagnetic fields the beams should never be physically influenced. Aharonov and Bohm, however, used quantum mechanics to show that there is between the two beams a phase difference proportional to the magnetic flux flowing inside the coil. Their surprising result was that "electrons can be physically influenced by electromagnetic fields without touching them."

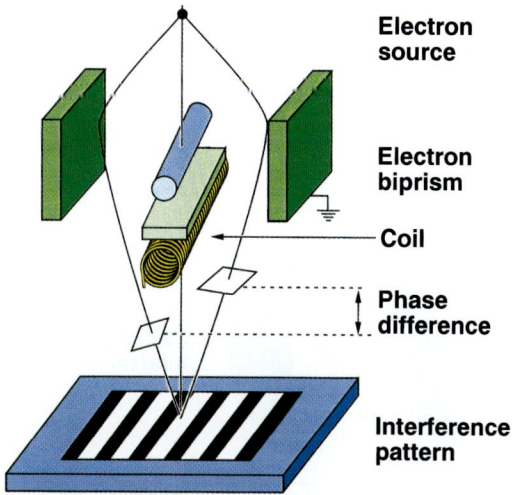

Fig. 68. The Aharonov–Bohm effect.
No magnetic fields are produced around an infinitely long coil where a current flows. Two electron beams starting from a source pass either side of the coil and are made to overlap. Although there are no magnetic fields along the two paths, a phase difference is produced between the two beams that is proportional to the magnetic flux inside the coil.

Public Reaction to the AB Effect

The AB effect attracted the attention of many physicists as soon as this paper was published, and most of them were initially skeptical. "Such a thing should never happen." Even famous physicists had never dreamt that electrons passing through field-free regions were affected by fields: R. E. Peierls, who was a referee of Aharonov's doctoral thesis, at first denied the AB effect but was convinced after having read the thesis. When Aharonov met him later, however, Peierls said that something was wrong. Richard Feynman was one of the exceptional physicists who accepted the AB effect at once. He sent a telegraph to Bohm and Aharonov as soon as the paper was published, "Congratulations! But I would like to have done it myself." The AB effect is in fact included in his 1963 textbook, *The Feynman Lectures on Physics.*

Chamber's Experiment

It was confirmation experiments that persuaded skeptical people. The effect was tested by such authorities in electron microscopy as Möllenstedt at Tübingen University, Boersch at the Technical University of Berlin, and L. L. Marton at the National Bureau of Standards in the United States.

The first result supporting the AB effect was reported in *Physical Review Letters* in 1960 by R. G. Chambers at Bristol University, who was not a specialist in electron beam techniques. He used a pointed iron *whisker* instead of a coil. A whisker is originally a long hairlike bristle growing near the mouth of a cat and certain other animals but here it means a single-crystalline needle. Magnetic flux flows inside this whisker, and leaks outside in the tapered region of the needle. This whisker was located just behind the filament of an electron biprism as illustrated in Fig. 69. Biprism interference fringes without a whisker are straight, as shown in Fig. 70(a).

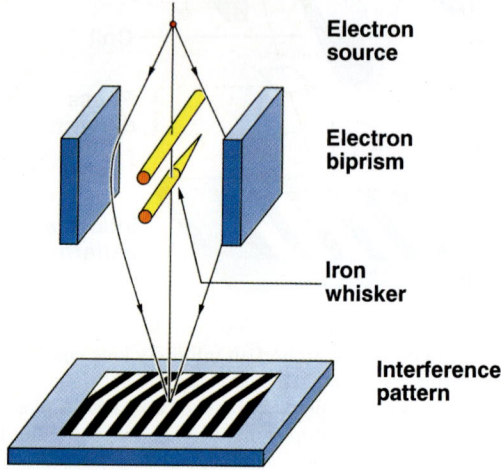

Fig. 69. Chambers' experiment.
        The phase difference between the two beams passing both sides of an iron whisker was proven to be produced proportional to the magnetic flux enclosed by the two beams.

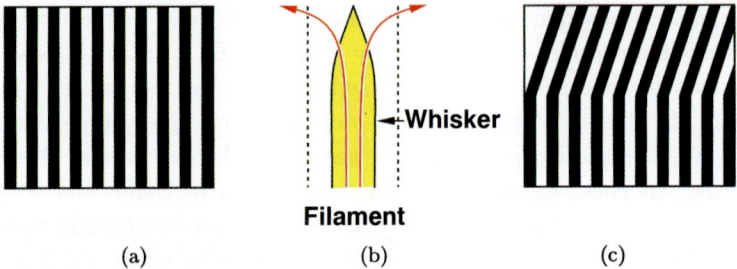

Fig. 70. Interference fringes in Chambers' experiment.
        (a) Parallel interference fringes without an iron whisker
        (b) Magnetic lines of force leaking outside of the whisker
        (c) Interference fringes when the whisker is located behind the biprism wire
        Since the phase difference between the two electron beams passing both sides of the whisker is given by the magnetic flux enclosed by the two beams, interference fringes are straight where the whisker is uniform in diameter, while the fringes are tilted near the tip of the whisker.

"What happens to the interference fringes when a whisker is placed in the shadow of the biprism filament as shown in Fig. 69?" According to the prediction of Aharonov and Bohm, a phase difference proportional to the magnetic flux enclosed by the two beams is produced between them. Therefore, interference fringes in the tapered region of the whisker, should be tilted as illustrated in Fig. 70(c), because the magnetic flux enclosed by the two electron beams passing on opposite sides of the filament decreases as the diameter of the whisker decreases. Chambers actually obtained such an interference pattern.

Möllenstedt's Experiment

The phase difference due to the AB effect was detected in Chambers' experiment and confirmed in a more elaborate experiment Möllenstedt and W. Bayh reported one year later. Möllenstedt and Bayh first developed a dedicated lathe machine for fabricating extremely fine coils, since the interference of electrons happen only in a narrow coherent region. As shown in Fig. 71, they fabricated coils with diameters ranging from 5 μm to 20 μm.

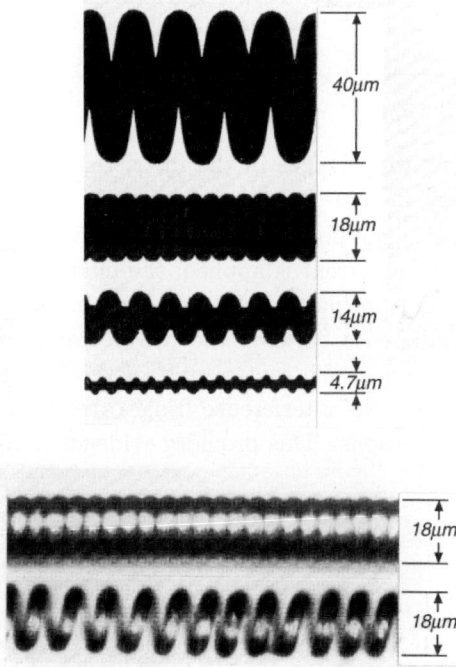

Fig. 71. Tiny coils for the AB effect experiment.
Möllenstedt and Bayh fabricated coils as thin as 10 μm or less with a lathe specially developed for testing the AB effect.

Möllenstedt and Bayh designed an experiment by which the dependence of a phase shift on the magnetic field in a coil could be recorded in one photograph (Fig. 72). Only a part of the interference pattern formed by passing electrons through a slit oriented perpendicular to the fringe direction was recorded on film, and the film was then made to move while an electric current applied to the coil

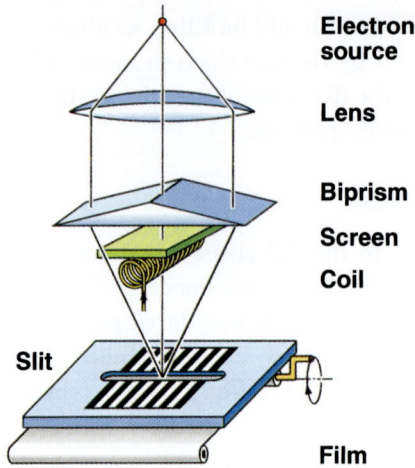

Fig. 72. Arrangement of Möllenstedt and Bayh's experiment.
    Due to the AB effect, interference fringes should be shifted when the current applied to
    the coil increases. Since only a slit region of the fringes is recorded on the film while it is
    moving, the dynamic behavior of the fringes can be observed in a photograph.

increased. In this way, the changes in the interference pattern corresponding to
changes in the magnetic field in the coil could be seen in one photograph.

In the photograph they obtained (Fig. 73), all the characteristics of the AB effect
can be observed: when no current is applied, the interference fringes are straight,
as can be seen in the bottom part of the photograph. When the applied current
increases, the phase difference between the two beams increases thus tilting the
interference fringes as seen in the central part of the picture. When the current
is fixed at a certain value, the interference fringes remain as they are and do not
return to the original positions. This provides evidence that the AB effect cannot

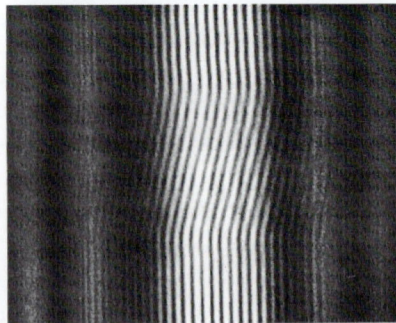

Fig. 73. Interference fringes obtained by Möllenstedt and Bayh.
    Interference fringes are straight when no current is applied (the lower part in the picture).
    When a current begins to increase, interference fringes are tilted (the middle part). When
    the current stops, the fringes do not return to the original position but remain displaced
    (the upper part), which indicates the AB effect.

be explained in classical terms using *electromagnetic induction*, but is a purely quantum-mechanical effect as Aharonov and Bohm asserted.

The AB Effect Controversy

In 1978, the AB effect suddenly became controversial. Of course, since its prediction, there had been much debate about the implications of the AB effect. But this time the situation was different. P. Bocchieri and A. Loinger in Italy asserted that the AB effect did not exist.

"What are the grounds of their assertion?" I will explain only the easy parts of their assertions: first, they asserted that vector potentials had no physical significance but were merely a convenient mathematical tool for solving electromagnetic problems. They claimed that vector potentials could be made zero everywhere outside the coil by selecting a special choice ("gauge") of vector potentials. They therefore concluded that the AB effect did not exist.

The second reason is concerned with the nature of the Schrödinger equation. In this equation the effect of a magnetic field is included in the form of vector potentials and not of magnetic fields, and for this reason it is easy to think that vector potentials have physical significance in quantum mechanics. However, they said that this was not the case, because the Schrödinger equation could be replaced by a set of equations, called *hydrodynamical equations*, using electric and magnetic fields only.

"How then could they interpret the experimental results by Chambers, Möllenstedt, and others?" S. M. Roy in India, for example, asserted that the experimental results could be attributed to the effect of the magnetic flux leaking outside from finite whiskers or coils. As for the first reason presented by Bocchieri and Loinger, it is true that vector potentials are not determined uniquely when magnetic fields are given, but the vector potential they proposed was pointed out not to correspond to the actual situation correctly, which will be described later in detail. Concerning the second reason, T. Takabayasi in Japan asserted that the effect of vector potentials in quantum mechanics could not be fully described by magnetic fields alone but required some additional boundary conditions.

Although many people were against Bocchieri and Loinger, they were not dissuaded. In four years the number of papers concerned with this controversy had reached 200. In 1980 we started to carry out confirmation experiments, but before going into these experiments here we should review the physical significance of vector potentials because their concept is vital to the interference of electron waves and also the interference of Cooper pairs in superconductors (which will appear in Chapter 12).

Chapter 10

VECTOR POTENTIALS, REAL OR NOT?

Aharonov and Bohm asserted that the AB effect was due to vector potentials. "What are vector potentials?" In the 1960s we learned at universities that vector potentials were merely a mathematical tool with which electromagnetic problems could be solved easily. Now, however, vector potentials are regarded as the most fundamental physical quantity in unified theories of all fundamental forces in nature. There, vector potentials are called *gauge fields*. In this chapter, I would like to give you a physical image of vector potentials and the history of vector potentials and then to tell you the story of our efforts until the physical reality or significance of vector potentials was established.

What are Vector Potentials?

Faraday introduced the concept of magnetic lines of force. The density of magnetic lines of force is called a *magnetic field*, or *magnetic flux density*. This quantity has not only a magnitude but also a direction. A physical quantity having both magnitude and direction is called a *vector* and is conventionally written with a bold and italic letter, such as B. The vector B has the direction of a magnetic line of force and has a magnitude proportional to the density of magnetic lines.

"What then are vector potentials?" The vector potential A is mathematically related to magnetic field B as follows:

$$B = \operatorname{rot} A . \tag{15}$$

If you are not familiar with mathematics, you may be puzzled at this equation. But please don't hesitate. It is only a representation of the relation between two quantities. Here "rot" is the abbreviation of rotation.

The meaning of this equation is as follows. If a distribution of A has a *rotational* component like a vortex, then the magnitude of B is given by the strength of the vortex and the direction of B is along a line perpendicular to the vortex plane. The vector B cannot yet be determined, since the line has two opposite directions. The direction of B is selected such that if you turn a right-handed screw in the vortex rotation direction, the direction in which the screw proceeds is the direction of B.

It is a matter of course that if there are no vortices in the distribution of A, the field B vanishes.

Such relations are often found in physics. The relation of an electric current i and the magnetic field B it produces is an example we already studied in Fig. 55(a):

$$i = \frac{1}{\mu_0} \text{ rot } B, \qquad (16)$$

where μ_0 is a constant. The electric current i produces a rotational stream of B around it.

Freedom of Vector Potentials

Once a magnet or a current-carrying coil is given, B is uniquely determined everywhere in a space. If not, we would be able to observe different distributions of magnetic lines of force using iron filings. However, B cannot be uniquely determined not only by Equation (16) alone. B can be determined by using also the fact that a stream of B has neither sources nor sinks, or magnetic lines of force have neither beginnings nor ends, or in mathematic terms div $B = 0$, where "div" is the abbreviation of divergence. For vector potentials, however, the situation is quite different. Since there are no such restrictions on A as div $B = 0$, vector potentials have thus an infinite number of freedom for a given arrangement: when A is a solution to Equation (15), $A' = A + \text{grad } \chi$ (χ: arbitrary function) also becomes a solution. Here "grad" is the abbreviation of gradient and the x-component of grad χ is given by $\frac{\partial \chi}{\partial x}$ and so on. The physical situation does not change by this *gauge transformation*: $A \to A + \text{grad } \chi$. You may ask "Can vector potentials be a physical quantity at all? Do many kinds of vector potentials lead to many results, thus making the physical law unpredictable?" Before answering these questions, we would first like to see examples of a variety of vector potentials.

Let us consider the simplest case of a uniform magnetic field [Fig. 74(a)]. Vector potential A for a uniform magnetic field is, for example, given by $A = \frac{1}{2} B \times r$, which is illustrated in Fig. 74(b). This is a *vortex flow* around a central axis determined by a magnetic line. The magnitude of A increases in proportion to the distance from the central axis. Since any magnetic line can be chosen as the axis of rotation, flows of vector potentials are quite arbitrary.

A completely different kind of vector-potential flow can be obtained by the gauge transformation: $\chi = \frac{Bxy}{2}$. This new vector potential A' is given by $A' = -By\hat{x}$ where $\hat{x}$ is unit vector in the x-direction. The resultant flow of A' is a laminar flow shown in Fig. 74(c). The vector potential A' vanishes in a certain vertical plane containing a magnetic line, and the value of A' changes in proportion to the distance from this reference plane. This reference plane can be replaced by a plane parallel-transported or rotated along the vertical magnetic line.

All these apparently different flows of vector potentials represent the same distribution of the uniform magnetic field shown in Fig. 74(a). There are many more

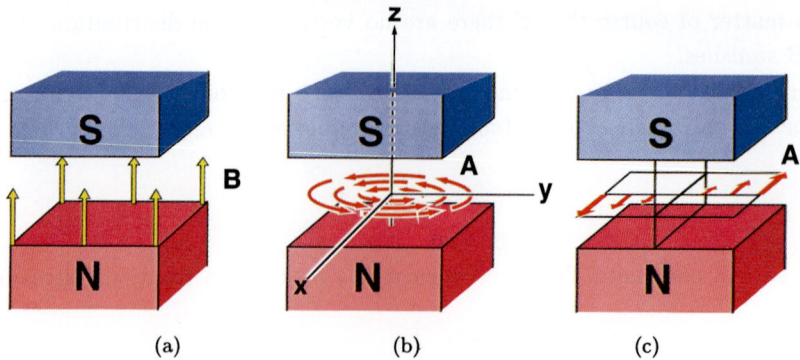

Fig. 74. Vector potentials for a uniform magnetic field.
(a) Uniform magnetic field
(b) Vector potential — vortex flow
(c) Vector potential — laminar flow

choices of vector potentials for this physical situation, and all the distributions of **A** represent a single physical arrangement. The different vector potentials can be changed from one to another by gauge transformations and are regarded as physically equivalent. You may not understand why vector potentials are so unintelligible. In order to help your understanding, I will show you later that vector potentials can be related to the number of magnetic lines of force having ever passed through at a point. The gauge freedom of vector potentials may be compared to the ambiguity in the number of the accumulated magnetic lines of force. It is shown later on that this freedom of **A** provided a key to the deeper truth of nature.

Electrotonic State

It was Lord Kelvin and Maxwell who introduced and developed the concept of vector potentials **A**. Here I would like to show you how the concept of vector potentials came to be devised.

The story begins with Faraday, who discovered *electromagnetic induction* in 1831: when a magnet comes close to a metal ring, a current is induced to flow in the ring in such a direction that the value of the magnetic flux flowing through the ring is kept unchanged [see Fig. 75(a) and (b)]. Faraday thought that the ring became in a "peculiar electrical condition of matter" when a magnet came close to it. The condition of the ring must have changed under the influence of the magnetic field, since a current was induced to flow in the ring when the condition was made to change from Fig. 75(a) to Fig. 75(b). Faraday called this state the "electrotonic state." Faraday later dispensed with this idea by means of considerations using magnetic lines of force: when magnetic lines of force crossed the ring wire, an electromotive force proportional to the change in the number of magnetic lines passing through the ring per unit time was produced, thus inducing a circulating current [see Fig. 75(b)].

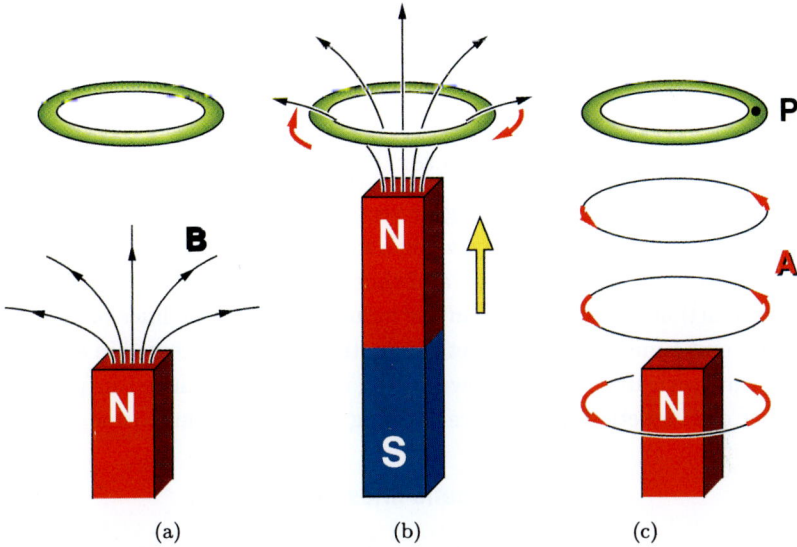

Fig. 75. Electromagnetic induction.
　　(a) A ring and a magnet
　　(b) A magnet approaching to a ring
　　(c) Vector potential A around the magnet
　　No current flows in a metal ring when neither the ring nor the magnet moves. When
　　the magnet comes close to the ring, a circulating current is induced in the ring so that
　　the magnetic flux passing through the ring may not increase but remain the same. This
　　phenomenon was discovered by Faraday and is called *electromagnetic induction*. Maxwell
　　later formulated this phenomenon using vector potential A. When the magnet comes
　　closer to the ring, A at point P in the ring increases, which, Maxwell thought, induces
　　electric field $E = -\frac{\partial A}{\partial t}$ to circulate a current in the ring.

　　　Maxwell attempted to mathematically describe Faraday's electrotonic state by
using the vector potentials A Lord Kelvin first introduced. Let us consider the
electromagnetic induction in terms of vector potentials A. Vector potentials are
circulating around the magnet as shown in Fig. 75(c). The more distant from
the magnet, the weaker the vector potentials. The magnitude of A at point P
on the ring increases when the magnet comes closer to the ring. Maxwell found
the electromotive force, or electric field E, produced proportional to the rate of A
changing with time. He expressed this in the following mathematical form:

$$E = -\frac{\partial A}{\partial t} \, , \qquad (17)$$

where $\frac{\partial A}{\partial t}$ means the change in A per unit time. As far as the magnet is coming
closer to the ring, $E = -\frac{\partial A}{\partial t}$ does not vanish. When the magnet stops moving,
the vector potential A at P no longer changes — *i.e.* $\frac{\partial A}{\partial t} = 0$ — and consequently
electric field E vanishes. The field E is induced only when vector potential A is
changing.

In this way, Faraday's electrotonic state was identified to be described by vector potentials. The next question is, "Can we connect magnetic lines of force and vector potentials?" Yes, the connection actually does exist!

Magnetic Lines of Force and Vector Potentials

Let us consider the physical meaning of vector potentials $\boldsymbol{A}$ in terms of magnetic lines of force. According to Faraday, when a magnet comes close to a metal ring, the electrotonic state in the ring changes and thus the electromotive force $\boldsymbol{E}$ is induced, which is proportional to the change in the number of magnetic lines passing through the ring per unit time. Electromagnetic induction can be expressed in a more simple case of a wire with unit length shown in Fig. 76: $\boldsymbol{E}$ is produced in proportion to the number of magnetic lines of force crossing the wire per unit time.

Magnetic lines of force

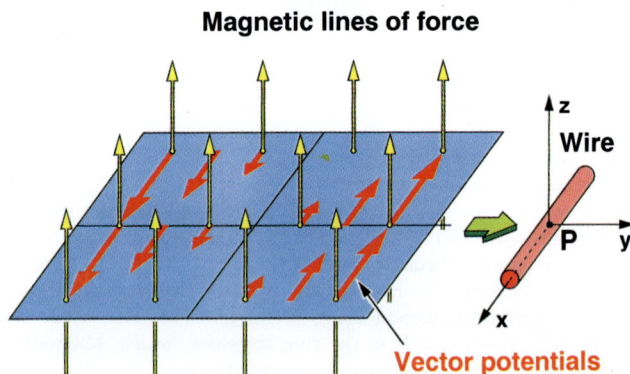

Fig. 76. What happens when magnetic lines cross a wire?
Faraday thought that electromotive force was induced proportional to the number of magnetic lines of force crossing the wire of unit length. While according to the Maxwell's way of thinking, electric field is produced proportional to the change in vector potential $\boldsymbol{A}$ at point P in the wire.

The two expressions by Faraday and Maxwell can be made equivalent if we interpret $\boldsymbol{A}$ as follows. *The magnitude of $\boldsymbol{A}$ is given by the total number of magnetic lines of force which have ever crossed the wire from left to right.* This interpretation is due to Whittaker. If you want to know more about it, please refer to his book, E. T. Whittaker, *A History of the Theories of Aether and Electricity* (Tomash/American Institute of Physics, 1987) Vol. 1, p. 200. Then the magnitude of $\frac{\partial \boldsymbol{A}}{\partial t}$ is equal to the number of magnetic lines of force crossing the wire per unit time.

Let us calculate the electromotive force using $\boldsymbol{E} = -\frac{\partial \boldsymbol{A}}{\partial t}$ for a concrete example of a uniform magnetic field crossing the wire shown in Fig. 76. A laminar-flow gauge of vector potential such that the direction of $\boldsymbol{A}$ is parallel to the wire is selected. When the magnetic field moves in the y-direction from left to right with a constant speed v, $\boldsymbol{A}$ is given by $\boldsymbol{A}(y, t) = -B(y - vt)\,\hat{\boldsymbol{x}}$. At point P $(y = 0)$, $\boldsymbol{A}(0, t) = Bvt\hat{\boldsymbol{x}}$.

Therefore $E = -\frac{\partial A}{\partial t} = -Bv\hat{x}$. The resultant electromotive force is given by Bv, which is the magnetic flux crossing the wire per unit time.

Now we can define A at point P more generally as follows: first we place a wire of unit length in various directions at point P and search for the direction for which the largest number of magnetic lines have ever crossed the wire. Then the direction of A is defined to be that direction and the magnitude of A is defined to be the total number of magnetic lines that ever crossed the wire. You may assert that there are ambiguities in counting magnetic lines of force. Your assertion is reasonable. The similar situation also exists in vector potentials. That is, vector potentials cannot be defined uniquely from Equation (15): $B = \text{rot } A$.

The physical image of vector potentials described up to this point is only for your intuitive understanding. We would like to consider more examples of vector potentials to understand more deeply what kind of physical effect vector potentials have on electron waves, since vector potentials play a key role in electron interferometry.

Vector Potentials Around a Coil

No magnetic fields exist around an infinitely long current-carrying coil, but vector potentials exist there and are circulating around the coil as shown in Fig. 77. This distribution of A may look similar to that corresponding to a uniform magnetic field shown in Fig. 74(b), but they are different in that with increasing distance from an axis the magnitude of A increases in the case of a uniform magnetic field but decreases in the present case of a coil.

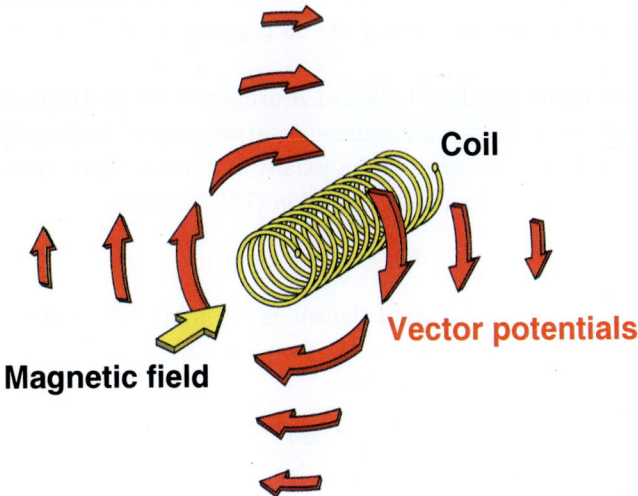

Fig. 77. The vector potential around an infinitely long coil.
No magnetic fields are produced outside the coil, but vector potentials exist outside of it because of the relation, $\oint A\,ds = \Phi$ [see Equation (18)].

"How is the distribution of A determined?" It can be derived from Equation (15): $B = \text{rot } A$ though not uniquely, but can be calculated more easily in the case of axially symmetric arrangements using a different equation equivalent to $B = \text{rot } A$:

$$\Phi = \int B \cdot \mathrm{d}S = \oint A \cdot \mathrm{d}s. \qquad (18)$$

Here, the *inner product* of two vectors — for example, $B \cdot \mathrm{d}S$ — appears in this equation. It is defined as the product of the magnitudes of the two vectors when the two vectors are in the same direction; but if the vectors are at angle θ, it is given by the product multiplied by $\cos \theta$.

Equation (18) is an *integral* form of the *differential* form $B = \text{rot } A$. The differential form gives the relation between B and A at a point, whereas the integral form gives the relation between A along an arbitrary loop and B inside the loop. The first term $\int B \cdot \mathrm{d}S$ is the integral of B over the surface enclosed by the closed loop. The integral is obtained as follows: we first divide the area into many small regions, and then in each region calculate the inner product of a small area $\mathrm{d}S$ and B there. "Why is the area represented by vector $\mathrm{d}S$?" It is mathematically convenient to represent the area by a vector whose value is given by the magnitude of the area and whose direction by the normal of the area. $\int B \cdot \mathrm{d}S$ is given by the sum of all these products. If B is uniform and perpendicular to the flat surface, then $\int B \cdot \mathrm{d}S = BS$, where S is the area of the surface bordered by the loop, and therefore is equal to the magnetic flux Φ passing through the loop.

The term $\oint A \cdot \mathrm{d}s$ in Equation (18) is the line integral of A along the closed loop. In this case also, we divide the loop into many line-segments and calculate the inner product of a line-segment $\mathrm{d}s$ and A. The line integral $\oint A \cdot \mathrm{d}s$ is given by the sum of the products for all the segments. If the direction of A is always along the loop and a constant, then $\oint A \cdot \mathrm{d}s$ is given by Al where l is the length of the loop.

We can easily derive the distribution of A using this integral form, Equation (18), especially in the case of an axially symmetric arrangements such as that in Fig. 77. If vector potential A is assumed to be axially symmetric and circulating along a circle with radius r around an axis, Equation (18) becomes

$$\Phi = \pi d^2 B = 2\pi r A, \qquad (19)$$

since the area of the cross section of the coil is given by πd^2 and the circumference of a circle whose radius r is given by $2\pi r$ (see Fig. 78).

Thus, A outside the coil is given by

$$A = \frac{B d^2}{2r}. \qquad (20)$$

The distribution of vector potentials outside an infinitely long coil is now clear: A outside the coil decreases with the distance r from the coil axis as illustrated in Fig. 79(a).

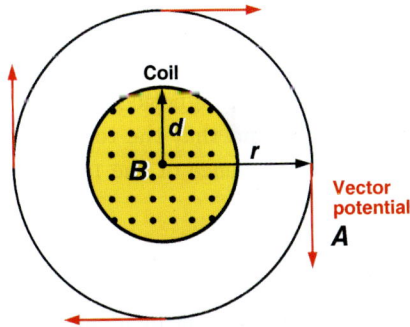

Fig. 78. Calculation of circulating vector potential outside the coil.
Since A is directed to the tangential line of the concentric circle r in radius, $\oint A \cdot ds$ is given by $2\pi r A$. Therefore $A = \frac{\Phi}{2\pi r} = \frac{Bd^2}{2r}$. The magnitude of A is in inverse proportion to r, the distance from the coil axis.

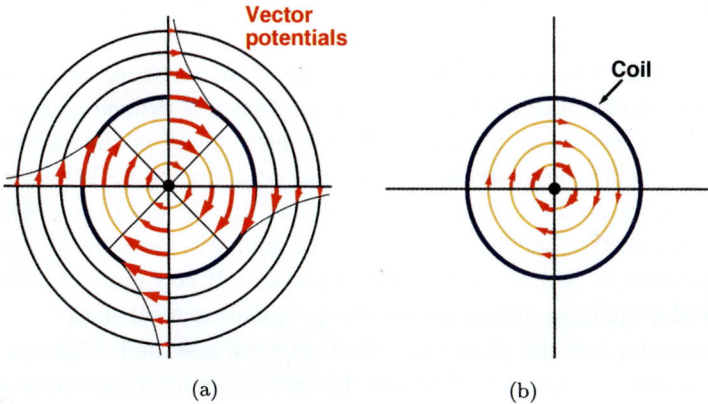

(a)   (b)

Fig. 79. Flow of A around the coil.
(a) Divergentless vector potential
(b) Vector potential vanishing outside the coil proposed by Bocchieri and Loinger

"What about vector potentials inside the coil?" The vector potentials can be obtained using Equation (19). In this case, the magnetic flux Φ enclosed by the closed loop is given by $B\pi r^2$ instead of $B\pi d^2$. Thus, A inside the coil is given by

$$A = \frac{Br}{2}. \tag{21}$$

The magnitude of A inside the coil increases in proportion to r as illustrated in Fig. 79(a).

Here I would like to touch upon a disputed point in the controversy over the AB effect. The distribution of A in Fig. 79(a) is continuous and its magnitude does not vanish even outside the coil. But as evidence for the nonexistence of the AB effect, Bocchieri and Loinger proposed such a vector potential illustrated in Fig. 79(b) that vanished outside the coil. This vector potential vanishes outside the coil but goes to infinity at the coil axis. Bohm and B. J. Hiley revealed that this vector potential corresponds to such a different physical situation that at the axis there is an infinitely thin coil which has the magnetic flux in the opposite direction, $-\Phi$, in addition to the magnetic flux Φ and that the total magnetic flux inside the coil vanishes. Therefore, it is the consequence of the negative flux at the central axis that $\boldsymbol{A}$ vanishes outside the coil.

Effect of Vector Potentials on Electron Waves

"What happens when an electron wave is incident on vector potentials?" The question is how electron wavefronts are displaced, or how electron waves are phase-shifted, when electrons pass through vector potentials. The interaction between an electron wave and vector potentials is determined by the Schrödinger equation:

$$\frac{1}{2m}\,(\boldsymbol{p} - q\boldsymbol{A})^2\,\Psi = i\hbar\,\frac{\partial \Psi}{\partial t}\,, \tag{22}$$

where $\boldsymbol{p}$ and q are the momentum and charge of an electron. Ψ is the electron wavefunction. In classical mechanics the momentum $\boldsymbol{p}$ is defined as $\boldsymbol{p} = m\boldsymbol{v}$ in the case of $\boldsymbol{A} = 0$ where m and $\boldsymbol{v}$ are the mass and velocity of the electron. In the case of $\boldsymbol{A} \neq 0$, $\boldsymbol{p}$ can be generalized such that $\boldsymbol{p} = m\boldsymbol{v} + q\boldsymbol{A}$. While in quantum mechanics $\boldsymbol{p}$ is replaced by an operator $-i\hbar$ grad, where "grad" is an abbreviation of gradient and means that the x-component of grad Ψ is given by $\frac{\partial \Psi}{\partial x}$ and so forth. Therefore, Equation (22) is a differential equation. If you want to know in more detail what the equation means, please study "quantum mechanics."

Let us consider how the phase of an electron wave is shifted. This can be known from the Schrödinger equation. The one-dimensional Schrödinger equation in case of $\boldsymbol{A} = 0$ is given by

$$-\frac{\hbar^2}{2m}\,\frac{\partial^2 \Psi(x,\,t)}{\partial x^2} = i\hbar\,\frac{\partial \Psi(x,\,t)}{\partial t}\,. \tag{23}$$

This wave equation can be solved for free electrons moving with constant speed v.

$$\Psi(x,\,t) = F\left\{\cos(kx - \omega t) + i\,\sin(kx - \omega t)\right\}, \tag{24}$$

where $k = \frac{2\pi mv}{h}$ and $\omega = E/\hbar$ (E: energy of the electron). This is a solution of the Schrödinger equation when $\boldsymbol{A} = 0$, and represents a plane wave which we have already discussed (Equation (24) becomes equal to Equation (9) by putting $t = 0$).

Let us consider the reason why the phase of the electron wave is not constant but changes spatially. When you look at the spatially changing phase term in this equation (*i.e.* $2\pi mvx/h$), the only factor responsible for the production of the phase shift is v. That means if $v = 0$ (electrons do not move), the electron phase is constant. Therefore, we can conclude that a spatial gradient of an electron's phase is produced because electrons are moving.

"Then, what about the influence of vector potentials A on an electron phase?" We already know that the spatial distribution of the electron phase is determined by momentum $p = mv$ in case of $A = 0$ [see Equation (24)]. That is, the spatial gradient of wavefunction Ψ is determined by mv in case of $A = 0$, but by $mv + qA$ in the case of $A \neq 0$.

Now the physical effect of vector potentials on an incident electron wave is evident: vector potential A produces a spatial gradient of an electron phase in a manner quite similar to that in which v does: electron velocity v produces a phase gradient along its traveling direction, while vector potential A produces an additional and independent phase gradient along the direction of A even if $v = 0$. It is a matter of course that a phase shift of an electron wave is not determined uniquely due to the gauge freedom of A.

Now let us consider the AB effect, using this knowledge of how A affects the electron phase.

The AB Effect and Vector Potentials

We know that no magnetic fields exist outside a current-carrying coil but vector potentials are circulating around it.

"What happens when an electron plane wave passes through both sides of the coil?" Since there are no magnetic fields, electrons pass straight through without being deflected. Their velocity remains the same. The effect of the electron velocity on the electron phase is given by Equation (24). The phase value increases with a constant rate along the direction of the electron motion. If electrons travel in a region *free of vector potentials*, the spacing of wavefronts, *i.e.* the wavelength, is equal.

However, since there exists vector potential A outside the coil, an additional phase shift is produced by it. The sign of the phase shift is opposite on both sides of the coil, or the wavefronts are displaced in the opposite directions as illustrated in Fig. 80. On the left-hand side of the coil, the vector potential A points downward as does v. Therefore, $|mv + qA|$ is smaller than mv, since $q < 0$. That means, the wavelength, *i.e.* the wavefront spacing given by $\lambda = h/p = h/|mv + qA|$ becomes larger there. On the right-hand side of the coil, the opposite thing happens and the wavefront spacing is shortened. Thus, the electron wavefronts on both sides of the coil proceed differently.

Let me express it more quantitatively. Since a different spatial gradient of the electron phase given by $\frac{2\pi q A(r)}{h}$ is produced at each point r, the total phase shift of an electron wave passing on one side of the coil is given by the summation of the inner product of $\frac{2\pi q A(r)}{h}$ and a path-segment dr, or by $\frac{2\pi q}{h} \int A(r) \cdot dr$. That is, a phase shift proportional to A integrated along an electron path is produced. This phase shift, however, cannot be observed. What we can observe is a difference in phase shifts between two electron waves starting from a single point, traveling two different paths, and arriving at another point as shown in Fig. 68. The resultant

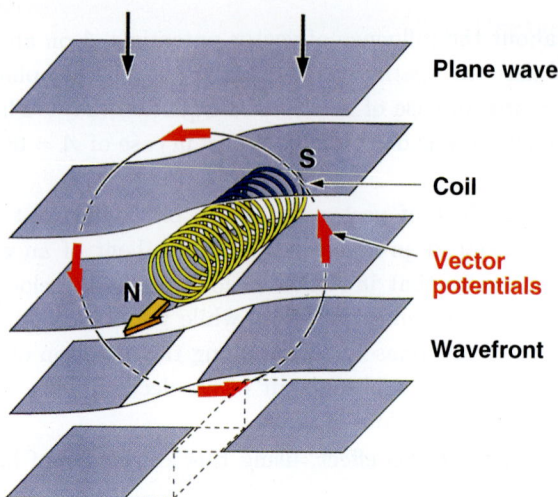

Fig. 80. Wavefronts of electron waves passing through both sides of the coil.
The phase of a traveling electron wave with speed v changes by 2π when the position is shifted in the propagation direction by $\lambda = h/mv$. Although an incident wave has wavefronts with an equal interval, when the wave enters the region where $\boldsymbol{A}$ exists, wavefronts are further advanced or retarded by $\boldsymbol{A}$ depending on which side of the coil they pass through. The total phase difference is given by $\frac{2\pi q}{h}\Phi$.

phase difference $\Delta\phi$ between the two waves passing through on the right-hand side and left-hand side of the coil is given by

$$\Delta\phi = \Delta\phi_R - \Delta\phi_L = \frac{2\pi q}{h}\left\{\int_R \boldsymbol{A}\cdot d\boldsymbol{r} - \int_L \boldsymbol{A}\cdot d\boldsymbol{r}\right\}.$$

You might think that $\Delta\phi$ vanishes since the values of these two integrals look the same. It is true that the paths of the two integrations both start from the same point A and end at the same point B. However, one path is on the right-hand side of the coil and the other is on the left-hand side. Therefore, they do not necessarily cancel each other.

The difference of the two integrals can be expressed more simply: since the integral changes sign when the starting point A and the ending point B are reversed: $\int_A^B \boldsymbol{A}\cdot d\boldsymbol{r} = -\int_B^A \boldsymbol{A}\cdot d\boldsymbol{r}$. Therefore, the difference of the two integrals is represented as $\oint \boldsymbol{A}\cdot d\boldsymbol{r}$ when the path of integration starts from the source point A, passes on the right-hand side of the coil, arrives at the image point B, then passes back on the left-hand side of the coil and returns to the source point A.

If we use the relation of $\boldsymbol{A}$ and $\boldsymbol{B}$ given by Equation (18), $\Delta\phi$ is expressed as

$$\Delta\phi = \Delta\phi_R - \Delta\phi_L = \frac{2\pi q}{h}\oint \boldsymbol{A}\cdot d\boldsymbol{r} = \frac{2\pi q}{h}\int \boldsymbol{B}\cdot d\boldsymbol{S} = \frac{2\pi q}{h}\Phi. \qquad (25)$$

In this way, a phase difference is proven to be produced between two electron waves traveling in regions free of magnetic fields on both sides of a current-carrying coil and to be proportional to the magnetic flux Φ enclosed by the two paths.

History of Vector Potentials

Vector potentials, about which we have so far learned a lot, have been an object of intense discussion for the past 100 years! Therefore, I will introduce here the history of vector potentials which owes much to an article by C. N. Yang [see C. N. Yang, "Vector Potential, Gauge Field and Connection on a Fiber Bundle" in *Quantum Coherence and Decoherence*, Proc. 5th Int. Symp. on Foundations of Quantum Mechanics in the Light of New Technology (ISQM Tokyo '95) Hatoyama, 1995, eds. K. Fujikawa and Y. A. Ono (Elsevier, Amsterdam, 1996) pp. 307–314.].

As I have already described, the story of vector potentials begins with Faraday, who established the concept of magnetic lines of force. This concept was mathematically formulated by Maxwell using vector potentials. The equations Maxwell obtained tell us that electric field E and magnetic field B are interrelated through vector potential A : E is produced when A changes with time $\left(E = -\frac{\partial A}{\partial t}\right)$, and B is produced when A changes spatially and has a vortex ($B = \text{rot } A$). Maxwell thought of vector potentials as a physical quantity, "momentum." The momentum of a particle is defined as the product of its mass m and velocity v. According to Newton's law of motion, the change in the momentum per unit time; *i.e.* $\frac{d\,(mv)}{dt}$, is equal to the force exerted on the particle. Since the force on a particle having a unit electric charge is given by $-\frac{\partial A}{\partial t}$, Maxwell thus conceived of A as an electromagnetic momentum.

The set of Maxwell equations thus established was not accepted quickly because the approach was so novel. It took thirty years before the Maxwell equations received recognition owing to independent efforts of O. Heaviside and H. Hertz. But they showed that the set of Maxwell equations could be written in terms of E and B only, and that A was not needed. Since that time A has been regarded as a mathematical auxiliary which has no direct physical meaning but is convenient for calculations.

In 1916, A. Einstein, who attemted to interpret gravitational forces by using the concept of fields in the same way Maxwell had explained electromagnetic forces, established the theory of general relativity where gravity was explained as the bending of space–time. He then directed his efforts to unifying both gravity and *electromagnetism*. The belief that a unified theory is possible did not originate with Einstein. A few tens of years before, Faraday had demonstrated that electric and magnetic phenomena, previously considered to be different kinds of phenomena, are closely related with each other. He showed that magnetic phenomena induced electric phenomena and vice versa. Faraday went so far as to think that even gravity might be related to electromagnetism, and he carried out experiments at the Royal Institution to test whether an electric current could be induced when a heavy object fell. The theory unifying electromagnetism and gravity is still one of the most important targets of today's physicists. It is amazing that Faraday's belief is still a theme in the forefront of physics.

Weyl's Vector Potentials

It was the mathematician H. Weyl who tried to realize Einstein's dream. In 1918 he tried to establish a theory unifying gravity and electromagnetism by extending Einstein's way of thinking that gravity can be explained using the geometry of space–time. According to Weyl, the gauge of length can be arbitrarily chosen at each point and the relation between the gauges at two neighboring points is specified by *vector potentials*. In other words, he explained electromagnetism in terms of structures of space–time. Einstein was impressed with his theory but rejected it: Einstein said that such a theory could not belong to physics, because a stick 1 m long could become 2 m long after going around a closed loop and returning to the original position. Although Weyl's attempt was unsuccessful, his way of thinking survives in modern *theories of gauge fields*.

Yang and Mills' Vector Potentials

The first step toward a unified theory was taken in 1954 by C. N. Yang and R. L. Mills, in whose work the unification is not that of electromagnetism and *gravity* but that of electromagnetism and *nuclear forces*. This paper, giving the basis of what is now called the *Yang–Mills theory*, is still one of the most frequently cited papers in the field of elementary particles. A more general theory was also constructed by R. Utiyama in Japan, though its publication was delayed. Weyl's fundamental way of thinking survives in these theories of gauge fields. Gauges of *length* Weyl thought to change from place to place are replaced by *phases* of an electron wavefunction. And the two phases at neighboring places are related with each other by electromagnetic *vector potentials*.

"How about Einstein's objection to Weyl in this case?" This time it is an electron instead of a stick that goes around a closed loop. No problem arises if the phase value at a point becomes different after one turn around a closed loop. Such phenomenon is none other than the Aharonov–Bohm effect.

In the *theories of gauge fields* — or in other words, the *theories of phase fields* — the fundamental principle is based on the assumption that *physics does not change when the phase of an electron wavefunction is locally changed*, since physics should not depend on the reference of the phase we can arbitrarily choose. Let us recall the wavefunction of a plane wave,

$$\Psi(x,\, t) = F\{\cos(kx - \omega t) + i\,\sin(kx - \omega t)\}. \tag{26}$$

where $k = \frac{2\pi m v}{h}$. The phase of this wavefunction is $kx - \omega t$. If the phase is changed by adding an arbitrary function $c(x,\, t)$, the wavefunction becomes

$$\Psi(x,\, t) = F[\cos\{kx - \omega t + c(x,\, t)\} + i\,\sin\{kx - \omega t + c(x,\, t)\}]. \tag{27}$$

The resultant wavefunction is no longer a plane wave. But this doesn't matter because there are no changes in the observable quantity, $|\Psi(x,\, t)|^2$. You might, however, when you think of the interference patterns we have often seen, wonder if the interference patterns will change under such a phase transformation.

An interference pattern is formed when two waves, starting from one point and tracing different paths, recombine at another point. It is a matter of course that if a phase difference is changed between the two interfering waves, the resultant interference pattern will change. But the phase transformation introduces only a common factor to the two interfering waves. This can easily be seen by rewriting Equation (27) as

$$\Psi(x, t) = F\{\cos(kx - \omega t) + i\sin(kx - \omega t)\}\{\cos c(x, t) + i\sin c(x, t)\}. \qquad (28)$$

The latter factor $\{\cos c(x, t) + i\sin c(x, t)\}$ is common to the two waves and consequently the interference pattern which is given by $|\Psi_1(x, t) + \Psi_2(x, t)|^2$ does not change at all.

Some of you might still doubt that the new wavefunction given by Equation (27) can be a solution of the original Schrödinger equation. It is true that the new wavefunction is no longer a solution of the original equation. However, we can make the Schrödinger equation invariant under the phase transformation making use of the gauge freedom of potentials: when we add an arbitrary phase $c(\boldsymbol{r}, t)$, we also have to change potentials, $\boldsymbol{A}$ and V, to new potentials, $\boldsymbol{A} + \frac{\hbar}{q}$ grad $c(\boldsymbol{r}, t)$ and $V - \frac{\hbar}{q}\frac{\partial c(\boldsymbol{r}, t)}{\partial t}$, which produces no physical changes. This transformation ($\Psi \to \Psi \exp\{ic(\boldsymbol{r}, t)\}$, $\boldsymbol{A} \to \boldsymbol{A} + \frac{\hbar}{q}$ grad $c(\boldsymbol{r}, t)$, $V \to V - \frac{\hbar}{q}\frac{\partial c(\boldsymbol{r}, t)}{\partial t}$) is called the *gauge transformation*.

Vector potentials, once discarded as rubbish, thus return to the stage as the main actor in the theories of gauge fields. Vector potentials are generalized and called *gauge fields*, which are regarded as the most fundamental physical quantity in these theories.

Theories of Gauge Fields

The Yang–Mills theory made the first step toward the realization of unified theories of fundamental forces. Indeed, the theory predicted three kinds of force. One corresponded to the electromagnetic force but the other two were different from nuclear forces in that they were long ranged like the electromagnetic. This turned out not to be due to a fault of the theory: it might be said that the underlying law of nature itself was true but its way of application was not appropriate.

We already know that the Schrödinger equation is invariant under a gauge transformation. Y. Nambu and G. Jona–Lasinio at Chicago University, however, found that even if the law itself was symmetric with respect to a transformation, there were cases where the symmetry of the solution was broken. An example of this *spontaneous symmetry breaking* will appear in the Section "Physical image of the wave of Cooper pairs" in Chapter 11, where a special gauge of vector potential is selected in a superconductor and therefore things are no more symmetric with respect to a gauge transformation. It was S. Weinberg and A. Salam who showed that electromagnetism and nuclear forces could be described in a unified form by theories of gauge fields: the spontaneous symmetry breaking provided a mechanism for changing the long-range forces appearing in the Yang–Mills theory to short-range forces.

The AB effect has become significant in theories of gauge fields, since it provides the only observable phenomenon directly indicating the physical reality of gauge fields.

AB Effect as a Theorem of Geometry

Einstein and Weyl attempted in vain to construct a unified theory using the concept of "geometry"; the theories of gauge fields at first did not seem to be related with geometry. In 1975, however, C. N. Yang showed that the theories of gauge fields could be described by an abstract geometry called a *fiber bundle theory*. The AB effect corresponds to a theorem in this geometry. We cannot study the fiber bundle theory here, but let us consider — just as a mental exercise — what this theorem is like.

Suppose an electron moves in a field-free space from point A to point B [see the upper illustration in Fig. 81(a)]. In an actual experiment, two electron beams starting from point A are recombined at point B by using an electron biprism. Let us

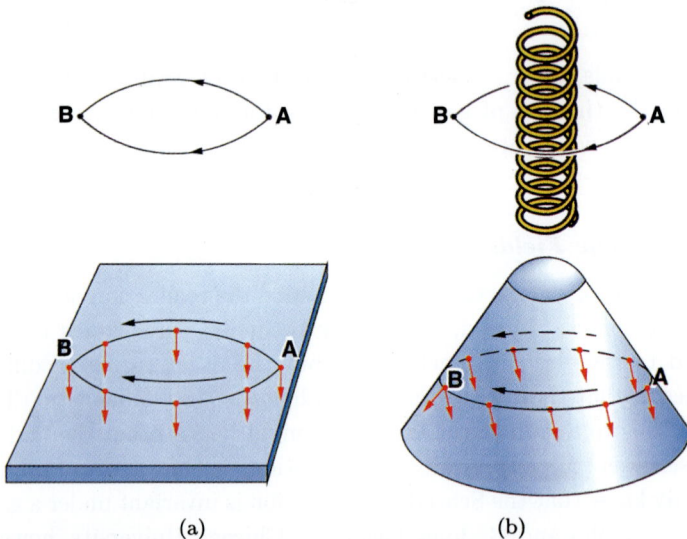

(a)   (b)

Fig. 81. Analogy between the AB effect and geometry in fiber bundle theory.
  (a) Two different paths in a field-free space without any magnetic field
  (b) Two different paths surrounding a magnetic flux in a coil
  The electron phase corresponds to the direction of an arrow. No phase difference is produced between the two paths traveling in a field-free space just as no change in the arrow direction is produced by parallel transports of the arrow in a flat plane as shown in (a). However, when two paths enclose a magnetic flux (a surface having a non-zero curvature), a difference in electron phase (in arrow direction) is produced although a magnetic field (a surface curvature) vanishes along the paths as shown in (b).

consider whether or not the electron phase at point B becomes different depending on the path the electron travels. Of course, the phase at point B depends on the path length, but we are concerned only with the additional phase if any. This situation corresponds to the following in the fiber bundle theory. A field-free space corresponds to a plane [see the lower illustration Fig. 81(a)], and the phase of an electron to the direction of an arrow in the plane. When the electron moves, the arrow is parallel-transported. An arrow starting from A and arriving at B has the same direction irrespective of the path it follows to get there.

Next, let us consider the geometry corresponding to the experiment of the AB effect. A field-free space outside the coil corresponds to a plane. The region inside the coil, where a uniform magnetic field exists, corresponds not to a plane but to a surface with a curvature proportional to the magnetic field. Since the magnetic field is uniform inside the coil, the surface becomes spherical. "How can these two surfaces be connected continuously?" The answer is shown in Fig. 81(b): a spherical surface is capped on the apex of a conical surface, since a conical surface is equivalent to a plane in that the conical surface can be unrolled to become a plane.

Now let us transport the arrow on the conical surface along two paths from point A to point B: one path goes on this side of the cone and the other on the back-side surface of the cone. If you think that it is difficult to parallel-transport the arrow on the conical surface, you can do so in an easier way. If you cut the conical surface, you can unroll it on a plane. There you can easily draw the parallel transport of the arrow. When you roll it again to the original cone, you will find the result at once: The two arrows oriented in the same direction at point A, which proceed along two paths, clockwise and counter-clockwise, are at point B oriented in directions different from each other, even though the arrows are always parallel-transported. This is a theorem in the fiber-bundle theory, and it corresponds to the AB effect. A phase shift is produced between two electron waves even though they move only in a field-free region.

Thus the AB effect became significant in the theories of gauge fields. Nevertheless, some people denied its existence, as described in the previous chapter. The dispute seemed destined to go on forever. For example, although three theoretical papers supporting the AB effect were published consecutively in *Physical Review* in 1981, the skeptics did not seem to be convinced. All the discussions until then had been theoretical, and proposals for elaborate new experiments began to appear.

Yang's Visit

We thought that the only way to end the controversy was to provide indisputable experimental evidence. All the previous experiments had used straight whiskers or coils with finite lengths which had N and S poles at both ends like the ones shown in Fig. 82(a). Since the essential point in the AB effect is that electrons passing

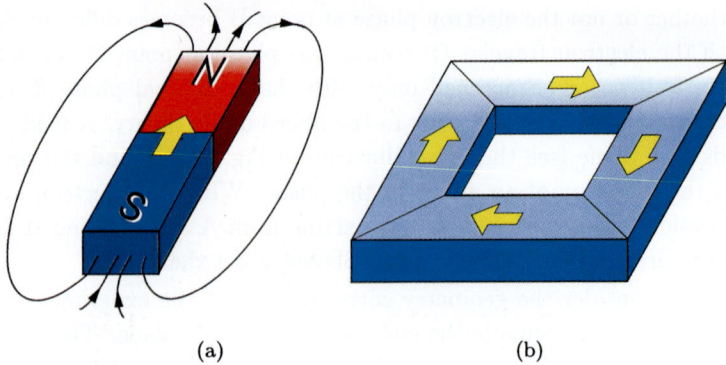

(a)   (b)

Fig. 82. Magnets.
   (a) Bar magnet
   (b) Toroidal magnet
   A straight bar magnet produces leakage flux unless it is infinitely long. While no flux
   leaks outside from a toroidal magnet even if it is finite.

through a *field-free* region can be physically influenced, we were determined to test
the AB effect under leakage-free conditions.

You might ask "Is it possible at all to make a leakage-free coil?" Yes, it is. The
principle is simple and straightforward. A magnetic flux cannot leak outside when
you connect four bar magnets without any gaps in order to form a closed magnetic
circuit [Fig. 82(b)].

The problem, however, was how to actually fabricate a tiny sample. Since the
electron wavelength is extremely short and the spatially coherent region is consequently small, a sample has to be made on an extremely small scale — say, less
than 10 μm — for the electron interference experiment. It seemed that it would be
difficult to fabricate such a sample unless a lithographic microfabrication technique
developed for making semiconductor devices was used. To do so, however, we had
to convince people that the experiment testing the AB effect is at least as important
as developing new devices is. I therefore wrote a letter to Yang, who made the first
great step to the theories of gauge fields, and indicated the significance of the AB
effect: "We are now planning experiments decisively testing the reality of the AB
effect. Are such efforts important for physics?"

One month later, at the beginning of June 1981, I received a telephone call
from the Physics Department in Tokyo University. It was from Yang, who in his
early 30s got the Nobel prize with T. D. Lee for his famous prediction of the parity
nonconservation! "I have just come to Japan and would like to visit your laboratory
for discussions." I had never expected such response, but he kindly visited our
laboratory at Kokubunji the next day (Fig. 83). I still believe that it was because
of his enthusiasm that we were able to continue our experiments long enough to
reach a clear conclusion.

Fig. 83. C. N. Yang (1922–).

New Experiments — 1

Soon after Yang's visit, we enthusiastically started the experiments in collaboration with the group on magnetic researchers. The experiments using toroidal magnets were completed in the following year, and they showed the reality of the AB effect! A phase difference was produced between an electron beam passing through the hole of a toroidal magnet and an electron beam passing outside the magnet (see Fig. 84). The phase differences detected agreed quantitatively with those predicted. To exclude the possibility that the detected phase difference might be produced by

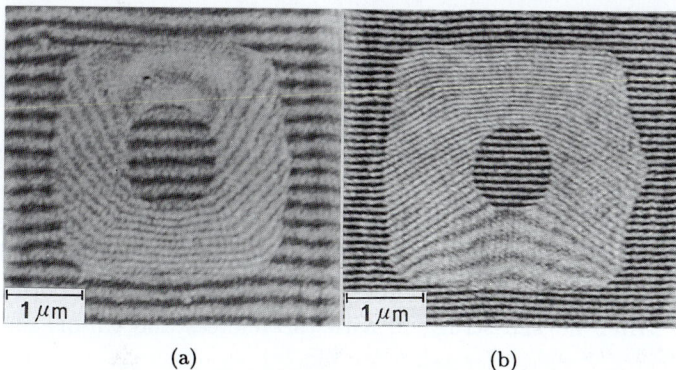

1 μm	1 μm
(a)	(b)

Fig. 84. Photographic evidence for the AB effect using toroidal magnets.
    (a) Clockwise rotation of magnetic flux
    (b) Counterclockwise rotation of magnetic flux
    A phase difference is produced between two electron beams having passed inside the hole and outside. The values of the phase differences in both cases, however, are reversed in sign.

some unwanted material remaining in the hole of the magnet, we measured the differences produced with two kinds of magnets: one whose magnetic flux rotate clockwise, and one whose flux rotates counterclockwise. If the phase shift were due to some substance, it would always be *advanced* inside the hole. The results shown in Figs. 84(a) and 84(b), however, indicated that the sign of phase shift depended on the direction of flux rotation. Naturally, we confirmed that leakage magnetic fluxes were too small to influence the conclusion. We did this by measuring them using interference electron microscopy. As shown in Fig. 85, if magnetic flux leaks from a magnet, we can see it at once.

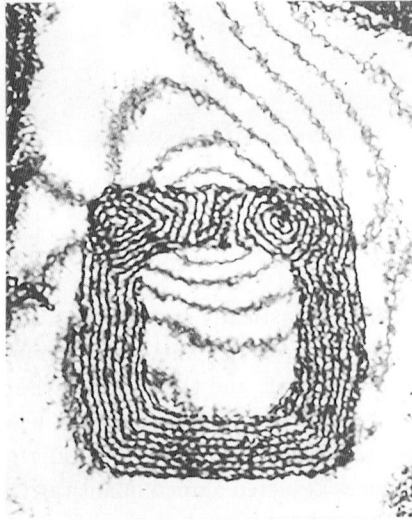

Fig. 85. An interference micrograph of a magnet having leakage fluxes (phase amplification: ×2). It can be seen from the micrograph at once that this toroidal sample has leakage magnetic fluxes.

Press Conference

We wrote up this result and submitted the paper to an American physics journal called *Physical Review Letters*, but within a month we were notified that the paper had been rejected. The opinions of the two referees were at the extremes: one said it was the first elegant demonstration of the AB effect; the other said it did not show the AB effect. The referees must have been selected from both sides of the controversy. After we had struggled with the referees and editor, the paper was published.

We then planned to hold a press conference and the *Asahi Shimbun* newspaper reported "An important theory verified — the conclusion to the 20-year controversy." Our result elicited a big response. I thought that this was thanks to the journal *Physical Review Letters*, and since that time I have been determined to have important results submitted to this journal.

More Experiments Demanded!

We expected this experiment to end the controversy, but some people remained un-convinced. One opinion published in an Italian journal of physics, *Nuovo Cimento*, was that "This experiment is not a confirmation experiment of the AB effect. When a part of an electron wave touches a toroidal magnet, the electron receives a force thus producing a phase difference. The observed result is due to the force and not due to the vector potentials."

We thought about writing a paper arguing against this interpretation, but be-cause I thought that this kind of experiment might resolve the 100-year dispute about the reality of vector potentials, I wanted to carry out such a decisive experi-ment so that all the theorists would be convinced.

Fortunately, an International Symposium on Foundations of Quantum Mechan-ics (ISQM) was first held at Hitachi Central Research Laboratory in 1983 and there Yang proposed a new kind of experiment. "If you cover a toroidal magnet with a superconductor, you can demonstrate the process of magnetic flux quantization dramatically."

He did not seem to imply that our experiment was not convincing. I took his proposal as his suggestion to carry out again a conclusive experiment of the AB effect under the perfect conditions.

New Experiments — 2

The proposed experiment looked extremely difficult. First of all, the fabrication of samples was not easy. I began asking U. Kawabe, a leader of our laboratory's research group on Josephson devices, for help in fabricating samples, and a year later he finally agreed to try it using the state-of-the-art lithography technology. The problem was not only making the samples. We also had to cool the samples to extremely low temperatures in our electron microscope. It was in the spring of 1985 that the preparations for the experiment were completed. Two years had passed since the start of developing a low-temperature sample stage.

The experiments did not go smoothly. Taking sample fabrication as an example, you might understand why simply by looking at the many processes required for it (Fig. 86). First of all, the design of the shape of toroidal magnets was a failure. You might have already noticed in Fig. 86 that we changed the magnet shape from a square toroid (Fig. 85) to a circular one to prevent the magnetic flux flowing inside the toroid from leaking outside. It was found, however, that almost all the magnets fabricated had leakage fluxes. A toroidal magnet did not form a completed magnetic circuit but was divided into several magnetic domains oriented in different directions producing the flux leakage.

The reason why a magnet divided into multiple domains were produced was soon clarified: The width of a magnet ring was too narrow compared with the ring radius to form a single domain. "Why was such a thin ring designed?" The microlithogra-phers insisted on taking into consideration the precision in the lithography that the

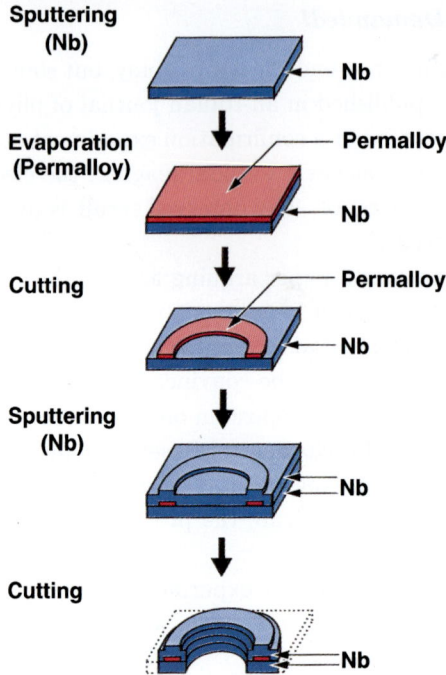

Fig. 86. Lithography process to fabricate toroidal magnets covered with superconductors.

diameter of a ring magnet be as large as possible and the ring width as narrow as possible so that a ring magnet made of permalloy could be completely covered with two sheets of superconducting niobium (see Fig. 86). They needed a margin large enough to enclose a ring magnet with the two niobium sheets. Since a ring magnet should not have multiple magnetic domains, we had to redesign the samples. First we confirmed that there was no leakage from the redesigned smaller magnets, and then we covered these magnets with superconductors. An example of such a sample is shown in Fig. 87. The sample is placed on a thin carbon film.

Such samples were set up in the microscope and cooled, but there were no signs that the niobium layers became superconductive. If the niobium layers become superconductive in a magnetic field, we should know it by observing a change in the deflections of incident electrons due to the exclusion of the magnetic field from inside the superconductor. We had to conclude that the samples were not cooled enough to become superconductive.

We had to redesign the samples again, and this could not be done soon. The microlithography requires many such processes as designing the pattern, making the photomask of the pattern, and finally fabricating the samples. This time ring magnets were self-supported by branches protruding from a niobium plate (Fig. 88). When we cooled the niobium plate by direct contact, samples connected to it would be cooled.

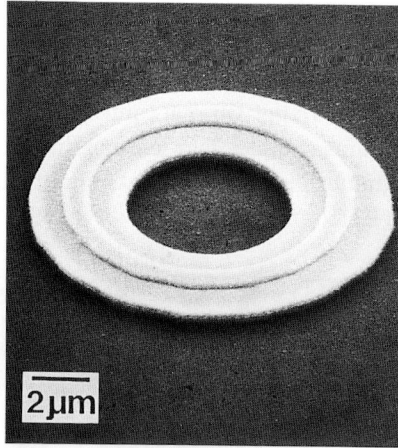

Fig. 87. Scanning electron micrograph of a toroidal sample.
   A sample was located on a carbon film, but could not be cooled enough to become super-conductive.

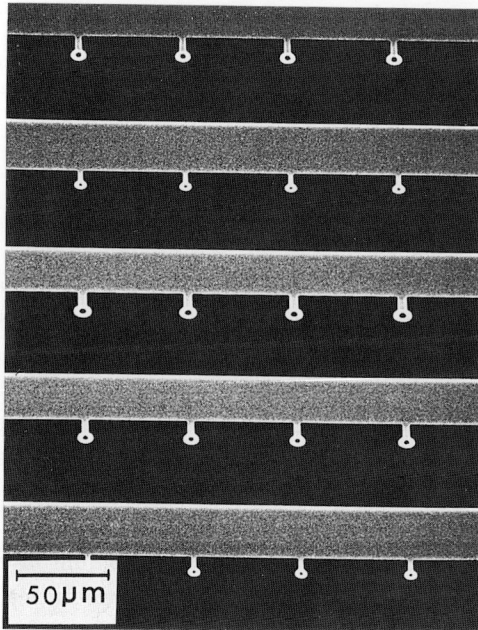

Fig. 88. Scanning electron micrographs of toroidal samples.
   In order to cool the samples to ultra-low temperatures, toroidal samples are connected to a niobium plate. Note that inner and outer diameters of the toroids are varied for the measurement of samples with various values of magnetic fluxes.

When you look carefully at the ring magnets in the photograph (Fig. 88), you can see that they have various outer and inner diameters. You can continuously change the magnetic flux of a coil by changing the current, but the flux of a magnet is determined by its geometry. To get various values of magnetic flux, we fabricated magnets of various sizes. Some samples had leakage fluxes, but such samples could be excluded because magnetic lines of force leaking outside the samples could be observed by interference electron microscopy (see Fig. 89).

Fig. 89. An interference micrograph of a toroidal sample with leakage fluxes.
Interference electron microscopy was used to check whether magnetic fluxes did not leak from the sample. This photograph shows an example of a sample with leakage magnetic fluxes.

Vector Potentials Restored!

A completed sample is shown in Fig. 90. Two niobium layers completely cover a ring magnet made of permalloy. To prevent even a slight leakage flux from the magnet, the niobium layer was more than 3000 Å thick, and any oxidized layer between the two layers was removed. A sample placed in the electron microscope was cooled to 5 K ($-268°C$) so that the niobium layers would become superconductive. The magnetic flux was confined within the superconductor layers due to the Meissner effect and did not leak outside. Evidence for this was obtained from the *quantization* of the electron phase difference, which will be explained in the section of flux quantization in Chapter 11. Incident electrons could not penetrate the magnet because of the covering of gold and niobium layers. The gold layer, which is not shown in Fig. 90(b), was formed by evaporating gold onto a sample from all directions to remove any electric field around a sample due to the contact potential differences produced between different materials.

The result is shown in Fig. 91. This time we cannot see interference fringes in the interior of the toroid image since incident electrons cannot penetrate the toroid. Interference fringes inside the hole and outside the toroid are displaced by just half a fringe spacing, that is, the phase difference is π. Since there are no magnetic fields outside the toroid, the fringe displacement must be produced by vector potentials.

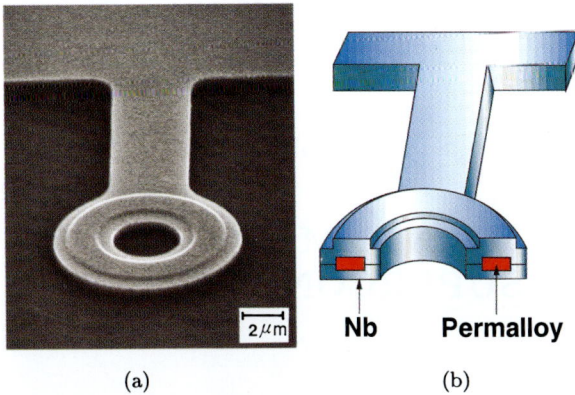

(a) (b)

Nb Permalloy

Fig. 90. A sample for testing the Aharonov–Bohm effect.
(a) Scanning electron micrograph of a toroidal sample
(b) Cross-sectional diagram of the sample
The bridge between the sample and the niobium plate is for cooling the sample.

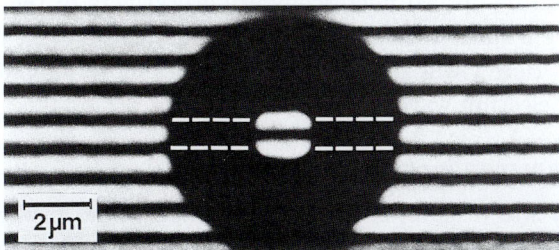

Fig. 91. Photographic evidence for the existence of the AB effect.
A fringe displacement inside the hole and outside the toroid indicates the existence of the
AB effect under the conditions where no magnetic fields leak from the sample.

We can conclude that this result provides crucial evidence for the existence of
the AB effect. It was thus clarified that vector potentials, which were regarded as
a mathematical tool, do produce an observable effect on electrons traveling in a
region free of magnetic fields.

Lastly I would like to add a word. Evidence that no magnetic fields leak outside the toroid is hidden in this photograph: the phase difference of π assures the
magnetic field inside the toroid is completely confined within the superconductor
by the Meissner effect. This is because the quantization of the phase in units of
π implies the magnetic flux quantization, which cannot occur unless the magnetic
field is completely enclosed by the superconductor.

Chapter 11

QUANTUM WORLD IN
SUPERCONDUCTORS

This last chapter will introduce you to the quantum world in *superconductors* which was unveiled by electron waves. When you hear of superconductors, you might imagine that everything inside them is completely frozen at ultralow temperatures, everything but supercurrents quietly flowing without any resistance. When you actually see the microscopic world in a superconductor, however, magnetic vortices too tiny to be seen by optical microscopy are everywhere in the superconductor, sometimes forming a static regular lattice and sometimes flowing either vigorously or intermittently. Here I will show you how vortices move in a superconductor, a scene revealed by electron waves.

Discovery of Superconductivity

Superconductivity was discovered by H. Kamerlingh Onnes in the Netherlands in 1911. Kamerlingh Onnes, who three years before had succeeded in liquifying helium, was investigating how the electric resistance of materials he purified changed when their temperatures approached absolute 0 K. Some people thought the resistance should approach zero at 0 K, others thought that it would become large because free electrons were frozen. The experimental results, however, confirmed neither of these predictions. Kamerlingh Onnes found that the resistance instead approached a constant value. He attributed this resistance to impurities and measured the resistance values of various kinds of the purest materials he could make. While measuring the resistance of mercury purified by an evaporation technique, Kamerlingh Onnes found that at 2 K the resistance suddenly disappeared. Soon he found that other materials, such as lead, showed a similar behavior that he called "superconductivity."

"Why did the resistance vanish?" The mystery became deeper as the properties of superconductors were further investigated and puzzled scientists for 50 years.

Unpractical Superconductors

Kamerlingh Onnes attempted to use superconductors practically, but when he applied a current, the superconducting state broke down before the current became high. The maximum current applicable, the *critical current*, was very low. He thought he might be able to make a powerful magnet by using a coil with many turns, but his attempt at this did not go well either: the superconducting state broke down when the strength of the magnetic field increased above a certain value (*critical magnetic field*). It was true that superconductors had no resistance, but the superconducting state was maintained only under extremely limited conditions. One had to realize that superconductors were far from practical applications.

New ways of practical applications, however, gradually opened up while the riddle of superconductivity was being resolved. Step by step, scientists learned that the mysterious properties of superconductivity were due to the rules of *quantum mechanics* governing the behavior of the electrons inside superconductors.

Magnetic Reactions of Superconductors

The properties of a superconductor to which a magnetic field is applied are really extraordinary. When the magnetic field is weak, the magnetic flux is excluded from the interior of the superconductor [Fig. 92(a)]. This is called the *Meissner effect* (after W. Meissner).

A circulating supercurrent flows at the surface layer of the superconductor, producing a magnetic field that cancels the magnetic field inside the superconductor. When the strength of the magnetic field reaches a critical value, the superconducting state is suddenly destroyed. This kind of superconductor is called a *Type I* superconductor, and both mercury and lead belong to this type.

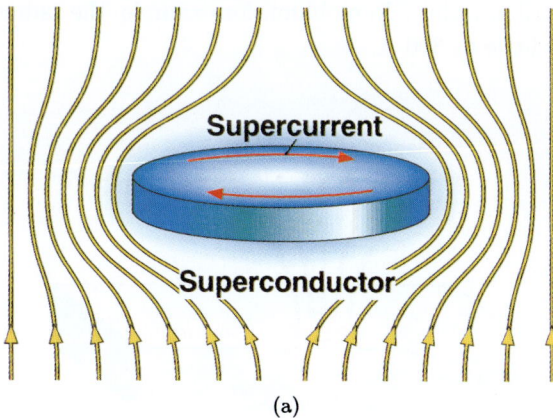

(a)

Fig. 92. Magnetic lines of force around a superconductor.
(a) Meissner state
(b) Vortex state
When a weak magnetic field is applied to a superconductor, the magnetic flux is excluded from the interior due to the Meissner effect. When the field increases, the superconducting state is destroyed in a Type I superconductor. While in a Type II superconductor, magnetic fluxes partially pass through it in the form of thin filaments called *vortices*.

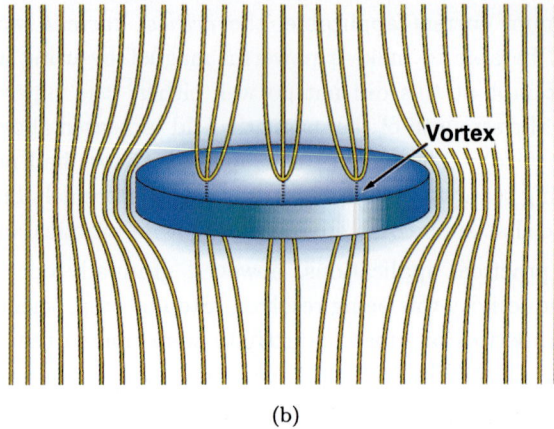

(b)

Fig. 92. (*Continued*)

Type II superconductors such as niobium were found later. In this type of superconductor a very ingenious mechanism exists: instead of breaking down the superconducting state, magnetic fluxes penetrate the superconductor in thin filament form [Fig. 92(b)]. Since magnetic lines of force making a long detour around the superconductor take a shortcut through it, the magnetic field near the surface of the superconductor weakens. Therefore, the superconductor can remain superconductive except the regions where the magnetic fluxes penetrated it.

A filament of magnetic flux is called a *vortex* and has a constant minute flux of $h/2e$ (2×10^{-15} Wb). A vortex not only has an extremely small flux value, but it is also extremely small in radius. In niobium, for example, the radius of the magnetic flux filament is as little as 300 Å.

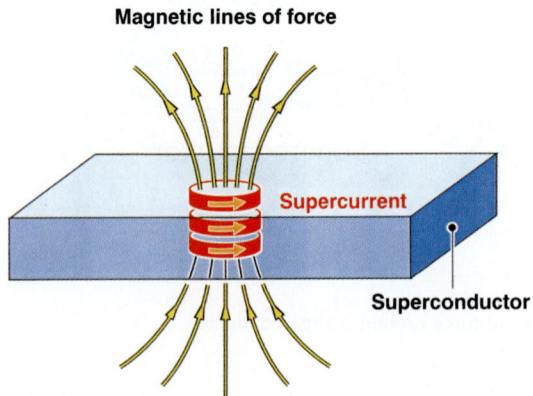

Fig. 93. A vortex.
   A vortex supercurrent is produced in a superconductor to form a thin filament of magnetic flux penetrating the superconductor.

"How can such a thin filament of magnetic flux be produced?" A vortex supercurrent flows forming a tiny electromagnet (Fig. 93). The vortex is also called a *flux line, a fluxoid,* or a *quantum flux.* As the applied magnetic field becomes stronger, the number of vortices increases in proportion to the magnetic field. Since the superconducting state breaks down partially in the regions inside vortices, the superconducting area decreases with an increase in the magnetic field, the superconductor is filled with vortices, and finally the superconducting state breaks down.

Observation of Vortices

It has been clarified that the superconducting state in a Type II superconductor can be maintained under a stronger magnetic field than in the case of a Type I superconductor by the production of vortices. However, when a current is applied to this superconductor, the superconducting state is again destroyed. This is because a Lorentz force proportional to the current acts on vortices (see Fig. 94). If vortices begin to move because of this force, a voltage difference is produced along the direction of the current by Faraday's law of electromagnetic induction. When a current flows with resistance, heat is produced. As a result of this heat, the superconducting state is destroyed in due time. In order to maintain the superconducting state even when a current is applied, vortices have to be fixed, or pinned down at some pinning centers against the Lorentz force. Tiny vortices hold a key to practical applications of superconductors.

"Can we observe vortices?" They are extremely tiny but play an important role in applications. The most successful early attempts to observe vortices used the Bitter technique, in which extremely tiny iron particles were sprinkled over the surface

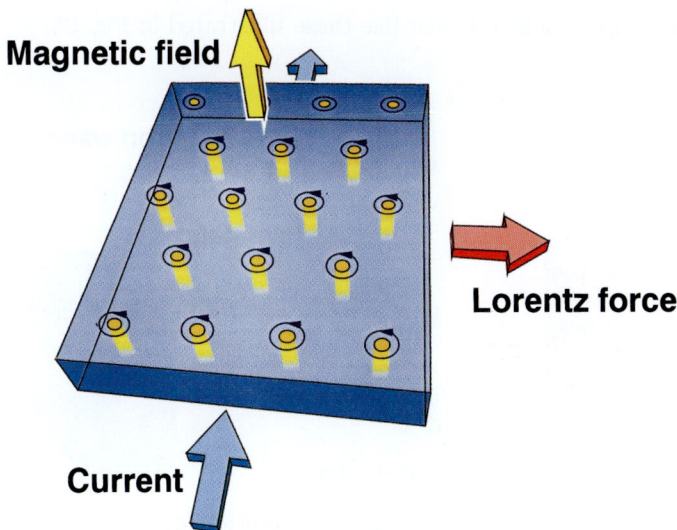

Fig. 94. Lorentz force exerted on vortices by current.

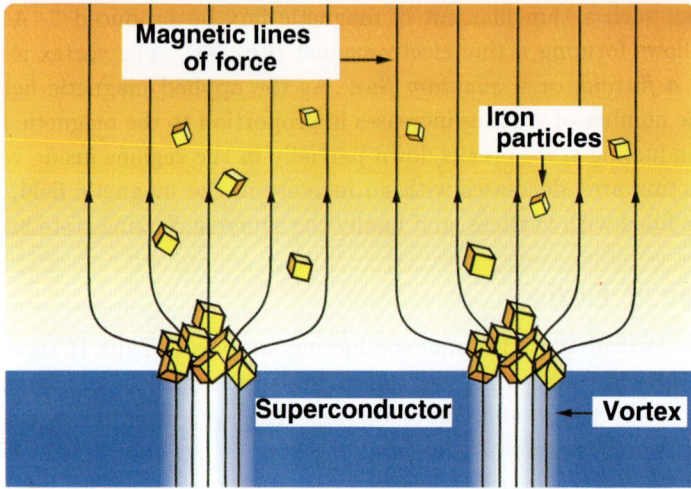

Fig. 95. Schematic diagram of the Bitter technique.

of a superconductor (see Fig. 95). The iron particles gathered at the locations where vortices appear on the surface are observed by electron microscopy (An example of a Bitter pattern appears in Fig. 110). Techniques developed recently use scanning tunneling microscopy (STM), a tiny Hall-effect device, a tiny SQUID device, and so on.

Magnetic Lines of Force of Vortices Observed!

We had been thinking for twenty years about using an electron beam to observe vortices, and in 1989 we were able to observe the magnetic lines of force leaking out of vortices in a superconductor, just like those illustrated in Fig. 95.

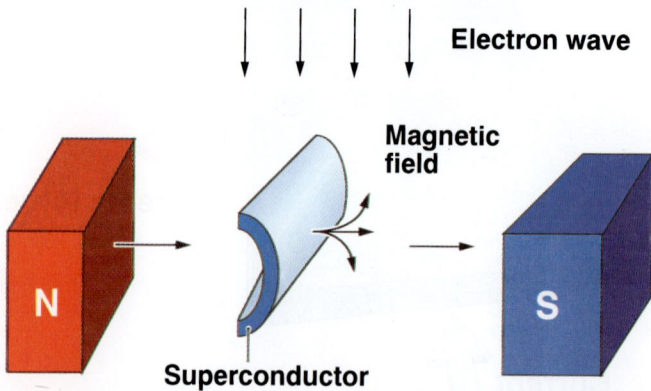

Fig. 96. A schematic diagram for observing magnetic lines of force of vortices by an electron beam. A magnetic field is applied perpendicular to lead thin film. Magnetic lines of force penetrating the film are observed by impinging an electron beam incident from above.

If we are to see the magnetic lines of force of vortices outside a superconductor, an electron beam has to be incident parallel to the surface of the superconductor and perpendicular to the magnetic lines. The experimental arrangement is shown in Fig. 96. A magnetic field perpendicular to a superconducting thin film is applied. A weak magnetic field cannot penetrate the film due to the Meissner effect. When the field becomes stronger, the magnetic field passes through the film in the form of vortices. An electron beam parallel to the surface is incident from above (see Fig. 96), and the magnetic lines of force are observed as an interference micrograph.

Fig. 97. Interference micrographs of magnetic lines of force of vortices leaking out of lead film
(phase amplification: ×2).
(a) Thickness = 0.2 μm
(b) Thickness = 1 μm

Figure 97(a) shows the result obtained when the sample is a lead film 0.2 μm thick. The lower black part indicates the film, and the upper part indicates a free space. Since this interference micrograph is phase-amplified twofold, one contour fringe corresponds to a flux quantum, $h/2e$. In the right hand side of the picture, you can see an isolated vortex: a magnetic line from a very small region of the superconductor surface fans out into space. Although the applied field is only a few gauss, the strength of the vortex magnetic field near the surface is as much as 1000 gauss. The magnetic field much stronger than the applied field is produced by the vortex supercurrent as illustrated in Fig. 93.

You might wonder whether the observed fringes show actually magnetic lines of force produced from vortices. That they are was confirmed by the following three observations. First, the magnetic lines of force observed remained existing even when the applied magnetic field was slowly removed. This proves that these magnetic lines of force were not due to the applied magnetic field. They must be due to the vortex supercurrent that could continue flowing even without any applied magnetic field, provided vortices were pinned at some defects. Second, when the temperature was raised above the critical temperature, the magnetic lines of force disappeared. This is the result of the sample changing into the normal state and the supercurrent stopping to flow. The third evidence was that the measured phase shift was exactly π, which corresponded to $h/2e$, the magnetic flux of a vortex.

An unexpected object was also found: a magnetic line forming a loop that can be seen in the left hand side of the picture in Fig. 97(a). This is a magnetic line produced from an antiparallel pair of vortices. Such pairs could not be detected when the Bitter technique was used because then the polarity of a vortex cannot be determined.

"How can a vortex directed opposite to the applied magnetic field be formed?" The Kosterlitz–Thouless theory predicts that a pair of opposite vortices will be created and annihilated in a thin film. The created vortex pair was thought to be frozen and pinned at some defects in the film. "Then what happens in a thicker film?"

Vortices in a Thick Film

An interference micrograph of magnetic lines of force penetrating a lead film 1 μm thick is shown in Fig. 97(b). The behavior of magnetic lines is quite different from that seen when the film is 0.2 μm thick shown in Fig. 97(a). Magnetic fluxes penetrate the film not as individual vortices but *in bundles of vortices*. No vortex pairs can be found.

This behavior can be interpreted as follows. Because lead is a Type I superconductor, when a strong magnetic field is applied to it the superconducting state breaks down at once.

In an actual sample, however, this does not usually happen. Even when the applied magnetic field is lower than the critical magnetic field B_c, the magnetic field becomes locally higher than B_c due to the Meissner effect, the superconducting state in the sample is destroyed partially, and the sample is divided into domains of normal and superconducting phases, which is called the "intermediate state."

An example is a case where a magnetic field is applied perpendicularly to a lead film shown in Fig. 97(b): the magnetic field applied to a film sample is not uniform at the sample surfaces but becomes high near the film edges as can be seen in Fig. 92(a). Therefore it can easily be understood that the superconducting state breaks down in the edge region at first. Normal and superconducting domains are arranged so that the total energy is minimal: in the case of a flat film uniform in thickness, the film is divided into parallel strips which are normal and superconductive alternately. Magnetic lines of force passing through normal regions were observed in Fig. 97(b).

According to the experimental results, however, the magnetic fluxes passing through normal regions correspond to an integral multiple of flux quantum as can be seen in Fig. 97(b). In other words, the fluxes can be counted by the number of vortices as 6, 3 and 4 from right to left in the photograph.

The flux quantization can be interpreted as follows. The film is not flat but curved as in Fig. 96. Due to this geometry of the film, normal regions are not strips but disks. Since the disk regions are surrounded by the superconducting region, magnetic fluxes passing through the disk regions are quantized in $h/2e$ units. An extremely thin film of lead behaves like a Type II superconductor, and magnetic fluxes penetrate the film in the form of individual vortices.

These vortices can also be observed even when they are moving. When the sample temperature was raised from 5 K, for example, the diameter of a vortex at the film surface became larger near the critical temperature T_c $(= 9.2$ K) and vortices began to move. Sometimes vortices oriented in the opposite direction penetrated the film from its edge and collided with original vortices to disappear together. Above T_c, all the vortices disappeared.

Protean Supercurrent

Interference electron microscopy has made it possible for the behavior of the magnetic lines of force of vortices to be observed. This technique has shown us that a *supercurrent* plays an important role in the magnetic behavior of a superconductor. Let us consider, for example, what happens when a magnetic field is applied to a superconductor and its strength is gradually increased. At first, magnetic fluxes are excluded from the interior of a superconductor by a supercurrent that flows in the surface layer and cancels the magnetic field inside the superconductor. When the applied field becomes stronger, vortices begin to appear inside the superconductor. These magnetic vortices are produced by vortex supercurrents, and their number increases in proportion to the strength of the applied magnetic field. The pictures in Fig. 97(a) are images of the magnetic lines of force of these vortices.

Supercurrents flow forever, without resistance and without producing heat. As soon as the applied magnetic field changes, the supercurrents respond by changing their magnitudes or their directions.

"What are supercurrents? How can they respond to the circumstances outside the superconductor?" Let us now consider this interesting and fundamental, but difficult, problem of superconductivity step by step.

Electromagnetic Induction

"Electromagnetic induction" discovered by Faraday in 1831 is in a sense similar to the Meissner effect. He found that when a magnet is brought close to a wire ring, a current flows in the ring. The current flows in such a direction that a magnetic flux passing through the space inside the ring does not change (This was already explained in Fig. 75).

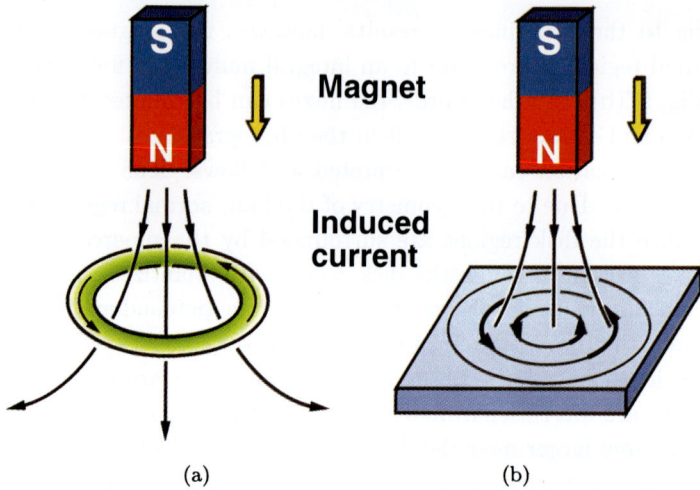

Fig. 98. Electromagnetic induction.
   (a) Metal ring
   (b) Metal plate
   A current is induced when a magnet is made to come close to a ring or a plate so that the magnetic flux enclosed by the ring or penetrating the plate may be unchanged. Therefore, a magnetic field is expelled inside the metal plate, but the Meissner effect cannot be explained merely by this electromagnetic induction.

This phenomenon occurs not only in a ring wire but also in a metal surface (see Fig. 98). The vortex current flows in such a way that the magnetic flux cannot penetrate the metal. The current flowing in the ring or at the metal surface soon weakens to vanish because of the nonzero resistance of the metal. The energy of the current is transformed into heat just as it is in an electric heater.

"What happens when the resistance vanishes? Can the Meissner effect be explained by the absence of resistance?" It may be true that a current once induced by electromagnetic induction is not damped and continues to flow, thus excluding the magnetic flux from the interior of the metal. But it is too soon for you to think you have successfully explained the Meissner effect in the metal with zero resistance. When a magnetic field is applied to it at room temperatures, a current will be induced but will die out due to the resistance. As a result, the magnetic field penetrates the metal without any disturbance. "What happens when the temperature is then decreased so that the resistance is zero?" Since the magnetic field has never changed, there is no room for electromagnetic induction. The Meissner effect thus cannot be explained by electromagnetic induction.

Diamagnetism

"How then can we explain the Meissner effect?" If we search for phenomena other than electromagnetic induction where a magnetic field applied to a material is prevented from penetrating it, we can find a phenomenon called *diamagnetism* though it is an extremely slight effect.

We just learned that when a magnet is brought close to a piece of metal, a circulating current is induced at the metal surface in such a way that the magnetic flux cannot ponetrate the metal. When the magnet stops moving, however, the current is soon damped. After that, the magnetic field passes through the metal without any disturbance, as if nothing had happened. But in some materials such as bismuth, a vortex current does not vanish completely and thus continues to repel the magnet. This is *diamagnetism* and is explained as follows.

Electrons orbiting an atomic nucleus is just like a metal ring as shown in Fig. 98(a). When a magnetic field is applied to these electrons, an additional circulating current is induced in such a direction that the applied magnetic field is suppressed. That is, the state of orbiting electrons changes. These electrons in an atom continue to flow forever without any resistance as long as the magnetic field is applied to it. To tell the truth, the explanation up to now in classical mechanics is not exact. Although a vortex current flows in each atom, the total current in a metal vanishes by cancellation. In quantum mechanics, however, the situation is different. A vortex current is quantized and the cancellation is incomplete, thus giving rise to diamagnetism. The effect of the diamagnetism in normal metals is too small to explain the Meissner effect. However, the fact is that a superconductor acts just like a single gigantic atom. Although it may be hard for you to believe that a macroscopic object follows the rule governing the microscopic world, this actually happens in superconductors! "If that's the case, can the Meissner effect be explained by the diamagnetic current?"

Gigantic Atom

The behaviors of superconductors are extraordinary and far beyond our experiences in this world: a supercurrent can flow eternally with zero resistance producing the Meissner effect and vortices. "Why and how can such things happen?" This greatest riddle in the present century was answered by J. Bardeen, L. Cooper and J. R. Schrieffer in 1957.

According to the theory of Bardeen–Cooper–Schrieffer, or the BCS theory, two electrons in a superconductor make up a pair called a *Cooper pair*. Although an electron is a Fermi particle, a Cooper pair composed of two electrons becomes a Bose particle. For this reason, all the Cooper pairs in a superconductor tend to be in the same state, forming a *single coherent wave*. We may say that a superconductor is really a gigantic atom!

There is, however, an essential difference between a superconductor and an atom. An atom has many seats for electrons which have different energy levels. Electrons are packed into these seats one by one from the seats with lower energy levels, since according to *Pauli's exclusion principle* multiple electrons cannot be in the same state. All the Cooper pairs, on the other hand, can be in the same lowest-energy state to form a single wave. If such a thing actually happens, it is no wonder that a superconductor shows *perfect diamagnetism*.

Let us begin with studying the properties of the wave of Cooper pairs.

Wave of Cooper Pairs in a Superconducting Lump

First consider how the wave of Cooper pairs behaves in an isolated lump of super-conductor. Let us consider the properties of the wave, keeping in mind that the wave is continuously connected everywhere inside the superconductor.

To begin with, we would like to recall how free electrons in a metal supercon-ductor behave at room temperatures. Free electrons move around in all directions. Therefore, the average velocity of electrons is zero. If it is not zero and electrons move in one direction on the average, a current must flow.

Let us think of such electrons as waves. A freely moving electron with velocity v can be considered to be a wave packet, which travels in the direction of the electron motion v and has wavefronts perpendicular to v. The average wavelength λ of the electron is given by de Broglie's relation, $\lambda = h/mv$ [Equation (1)], from which we know that slower electrons have longer wavelengths.

"What will happen to such electron waves when the superconductor is cooled to become superconductive?" You might think that as the superconductor temper-ature decreases, electrons slow down until they stop moving. This is not the case. The speed of free electrons hardly changes. Only such electrons that move with high speed slow down slightly.

Some of you may wonder how free electrons can move if not with thermal energy. The reason why free electrons move with speed much higher than that of thermal motion comes from the quantum-mechanical effect. Just like in the case of an atom, there are many seats with different energy levels for free electrons in a metal. Free electrons are packed into the seats from lower energy levels. Electrons seated in the highest energy level have to move with speed as high as 100 km per second! These electrons cannot slow down simply because the lower energy levels are packed and also because electrons cannot share the lower-energy seats since electrons are Fermi particles. They have to continue moving even at 0 K. Only such high-speed electrons form a Cooper pair.

"In what directions do free electrons move?" The direction of the electron movement is determined by the seat of electrons. Some electrons move in the right direction and others in the left direction, and the average velocity of free electrons is zero. This is a world which cannot be understood by common sense. If you want to understand it, please proceed to study "quantum mechanics."

If there is a free electron moving with velocity v, there is always a free electron moving with the opposite velocity $-v$. These electrons form a Cooper pair when the superconductor is cooled below T_c to become superconductive. Therefore, the velocity of a pair is zero, which according to de Broglie's relation means that the wavelength of the wave of a Cooper pair is infinitely long. It should be noted here that Equation (1) is valid not only for electrons, but also for Cooper pairs when the mass and speed of Cooper pairs are used.

"What happens to Cooper pairs when the wavelength becomes infinitely long? Are they no longer waves?" They remain as waves. If the wave has a finite wavelength, a phase shift of 2π is produced between two points that are one-wavelength apart in the direction of the wave propagation. But the wavelength is now infinitely long, and therefore the phase does not change anywhere in the superconductor. (If the phase changes spatially, the wave would have to propagate in the direction of the phase change.)

Cooper pairs are Bose particles and therefore all the Cooper pairs can occupy a single state: all the waves of Cooper pairs are connected to form a single coherent wave by making the phases of all the waves the same.

In this way, a coherent wave of Cooper pairs is produced in an isolated lump of superconducting material and this wave has a phase value that is constant everywhere in the superconductor. This is the simplest form of a wave, but may be difficult to understand since its spatial distribution does not look like that of a wave. This corresponds to the case of $k = 0$ in Equation (25): $y(x, t) = F\{\cos(kx - \omega t) + i \sin(kx - \omega t)\}$. Even in this case the wave does oscillate: the oscillation takes place simultaneously at every point.

Phase Changes When Electrons Move

The phase of Cooper pairs turned out to be a constant inside a lump of superconductor when no magnetic field or current is applied to it. You might wonder when and how Cooper pairs are phase-shifted. Let us consider this problem using the case of an electron beam instead of Cooper pairs, since we have already had some knowledge about an electron beam. The ways that phase shifts are produced must be similar for both cases, since both waves follow the same Schrödinger equation. The solution of the Schrödinger equation has already been obtained for an electron plane wave in Chapter 10:

$$\Psi = F\{\cos(kx - \omega t) + i \sin(kx - \omega t)\}, \qquad (29)$$

where $k = \frac{2\pi m v}{h}$. The phase of this wave is not a constant in space, but is given by $kx - \omega t$.

Since $k = \frac{2\pi}{\lambda} = \frac{2\pi m v}{h}$, the spatial gradient in the phase is produced because electrons are moving.

Equation (29) represents the wavefunction for an *electron* plane wave, but also becomes the wavefunction for a plane wave of *Cooper pairs* if the values of m and v are replaced by the values for Cooper pairs. When Cooper pairs do not move (*i.e.* $k = v = 0$), then Equation (29) represents the wavefunction of a spatially constant phase [see Fig. 99(a)].

"How can we actually produce a wave of Cooper pairs with phase changing spatially?" We have only to move Cooper pairs in one direction. If Cooper pairs move with speed v, then a phase gradient represented by Equation (29) must be produced (see Fig. 99).

A phase gradient in the wave of Cooper pairs is produced not only when Cooper pairs move. The same thing happens when there exists vector potential $\boldsymbol{A}$.

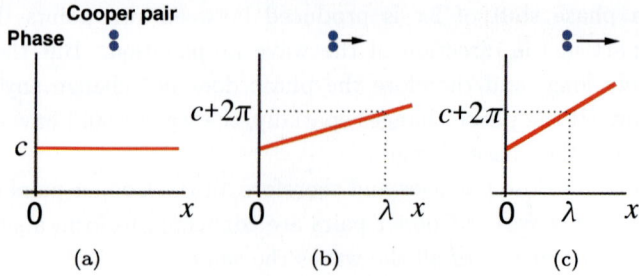

Fig. 99. Spatial gradient in the electron phase produced by Cooper-pair movement.
(a) No gradient produced when Cooper pairs do not move
(b) A gradient produced when Cooper pairs move
(c) A steeper gradient produced when the speed increases

One More Factor Producing Phase Shift

Vector Potentials

"How does an electron wave behave in a magnetic field?" When there is no magnetic field, a phase gradient given by $k = \frac{2\pi m v}{h}$ is produced when electrons move with velocity v. When there is a magnetic field, a phase gradient of

$$k = \frac{2\pi m}{h} \left(v + \frac{q}{m} A \right) \tag{16}$$

is produced.

We have finally arrived at the point where we know how the wave of Cooper pairs is phase-shifted. Now we want to see, using the knowledge obtained, how the Meissner effect is produced. First, I'll summarize the effect here: a magnetic field is applied to a superconductor whose temperature is above its T_c. When the superconductor is then cooled and the temperature falls below T_c, it becomes superconductive and excludes magnetic flux from its interior. This cannot be explained by electromagnetic induction. "Why does magnetic flux have to be excluded from the superconductor?" Let us first consider what would happen to the wave of Cooper pairs if an applied magnetic field should penetrate a superconductor.

What would Happen if a Magnetic Field should Penetrate a Superconductor?

"Is it possible at all to maintain the superconducting state when an applied magnetic field passed through a superconductor?" — I will show you that if a magnetic field passed through a superconductor, the wave of Cooper pairs cannot exist, since the phase at a point cannot be determined uniquely.

Let us consider the phase of the Cooper pair after one turn starting from point A, tracing a circular path, and returning to point A (see Fig. 100)? If the phase does not return to the original value, the wave of Cooper pairs cannot exist. This condition may be too stringent. The wave of Cooper pairs can also exist even when

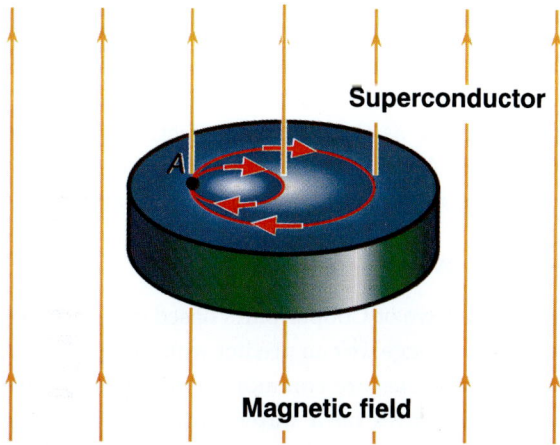

Fig. 100. Phase difference after one turn along a closed loop.
Since the phase difference is determined by the magnetic flux enclosed by the loop, the wave dose not persist unless the phase shift after one turn is $2n\pi$ (*n*: integer).

the phase difference is an integral multiple of 2π. This case actually happens in magnetic flux quantization, which will appear later.

You may remember that the phase difference between two *electron beams* starting from a point, tracing different paths, and ending at another point is given by the product of $2\pi q/h$ and the magnetic flux enclosed by the closed loop connecting these two paths [see Equation (25)]. The situation for Cooper pairs is completely the same as that for electron beams. The phase difference produced when Cooper pairs make a trip around a closed loop is $2\pi q/h$ times the magnetic flux enclosed, where q is now two times the electric charge of an electron.

If little by little we make the circle larger, the magnetic flux enclosed by the circle — and consequently the phase difference — increases gradually as a result of the magnetic field penetrating the superconductor. Therefore, there must be a case in which the phase value does not return to the original value. We must thus conclude that the wave of Cooper pairs could not exist when a magnetic field should penetrate a superconductor.

Please recall here that a spatial gradient in the phase distribution is given by $\mathbf{k} = \frac{2\pi m}{h}\left(\mathbf{v} + \frac{q}{m}\,\mathbf{A}\right)$: The phase gradient is caused not only by vector potentials $\mathbf{A}$ but by the movement $\mathbf{v}$ of Cooper pairs. Therefore, you may wonder if the wave of Cooper pairs can exist by the effect of $\mathbf{v}$ even when a magnetic field penetrates a superconductor. Such a thing actually happens in special cases where the magnetic field passes through it partially. For example, when a magnetic field penetrates a surface layer of a superconductor, the superconducting state can be maintained even in the layer by the effect of $\mathbf{v}$ (Meissner effect). However, when a magnetic field penetrates the *whole* superconductor undisturbed, the superconducting state has no choice but to break down.

We have seen that the wave of Cooper pairs disappears when a magnetic field penetrates a superconductor. The superconducting state can be maintained only when a magnetic field is prevented from penetrating the interior of a superconductor. This actually takes place as the Meissner effect. "By what mechanism is the Meissner effect produced?" Although we cannot easily come close to the answer, we first need to know the properties of the wave of Cooper pairs here before proceeding.

Physical Image of the Wave of Cooper Pairs

Let us get an image of the wave of Cooper pairs based on the material we've learned up to now. Once we get the image we can predict what happens even when situations change, since the wave properties are common to all kinds of waves.

You can easily imagine waves on a water surface. The wave of Cooper pairs in an isolated lump of superconductor is a wave of constant phase. This wave can be compared to a water wave whose surface is flat but moving up and down as a whole. The phase of the water wave is constant independent of the location and the wavelength is infinitely long.

"What happens when a current is applied to a superconductor?" A supercurrent flows, and the wave of Cooper pairs propagates in the direction of the velocity v of Cooper pairs, its wavefronts being perpendicular to v. Since the wavelength λ of the wave is given by h/mv, a spatial gradient of the phase is produced such that the phase changes by 2π at intervals of λ in the direction of the steepest phase change.

A spatial gradient of the phase can be produced by vector potentials A even if $v = 0$. Wavefronts then become perpendicular to the direction of A and propagates in the A direction. Finally, if $A \neq 0$ and $v \neq 0$, the wavefronts are perpendicular to the direction of $v + \frac{q}{m} A$ and propagate in the direction of $v + \frac{q}{m} A$.

We could thus get some image of the wave of Cooper pairs even when a magnetic field or a current was applied. It was also clarified that the behaviors of the wave were determined by the quantity,

$$u = v + \frac{q}{m} A. \tag{31}$$

This quantity u is proportional to the generalized momentum $p = mv + qA$ which appeared in the Schrödinger equation [Equation (22)]. Consequently p is the total momentum of a charged particle including the effect of A. Please recall that Maxwell regarded A as an electromagnetic momentum.

Some of you may not be satisfied with this physical image of u. Let us try to interpret u more intuitively using vector potential A which we grasped in terms of magnetic lines of force discussed in Fig. 76.

It can be interpreted that the magnitude of vector potential A at a point is proportional to the number of magnetic lines of force having ever passed across a wire of unit length located at the point from left to right when the wire direction is selected so that the number becomes the largest (see Fig. 76). Suppose that the wire is in the superconducting state and therefore has no resistance, then a supercurrent is induced and accumulated every time magnetic lines of force cross the wire. Of course in order for a supercurrent to flow, both ends of the wire have to be connected to form a superconducting loop. Since A is proportional to the number of magnetic lines of force having crossed the wire so far, vector potential A at a point may be interpreted to be the supercurrent induced so far in the virtual wire due to magnetic lines of force having crossed the wire. It can be interpreted that u given by Equation (31) indicates the total flow which is given by the summation of this virtual flow $\frac{q}{m} A$ and the real flow v of supercurrent.

Properties of the Flow of the Wave of Cooper Pairs

Now we have arrived at the conclusion that u indicates something like a velocity of the total supercurrent including the effect of electromagnetic fields. Then there must be some kind of flow represented by the velocity field u. Let us consider what this flow of u is like and what properties the flow of u has. The flow of v, represented by the first term in Equation (31), is easy to understand. Since this is the flow of actual charged particles of Cooper pairs, it is like a river stream and has neither sources nor sinks. The flow of vector potentials A, given by the second term of Equation (31), has a variety of choices for a given distribution of magnetic field B because of the freedom of the gauge transformation. We can, however, select a flow that has neither sources nor sinks. Then, we arrive at the conclusion that the flow of u, which is the summation of two kinds of flows, v and A, has also neither sources nor sinks.

This vector u, or $k = \frac{2\pi m}{h} u$, gives a spatial gradient to the phase of the wave of Cooper pairs. This flow is like a stream of water flowing down a mountain whose height is given by the phase value (see Fig. 101), except that the flow of u is actually in the opposite direction — from lower potentials to higher potentials. Since we would like to compare this flow to the water flow, think that the mountain height in Fig. 101 represents not the phase value itself but the value multipled by -1.

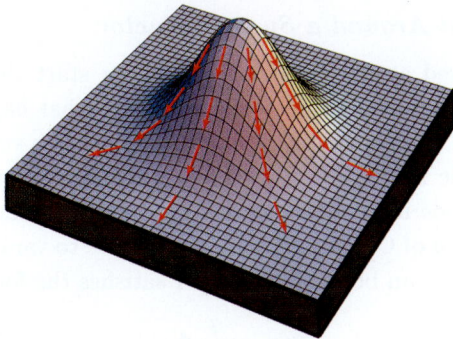

Fig. 101. Water streams flowing down the mountain surface.
　　　　The flow of the wave of Cooper pairs is similar to a water stream. The Cooper-pair wave flows down the distribution of the phase. The truth is that the flow u is directed from lower places to higher places of the phase distribution.

"Can the wave of Cooper pairs flow everywhere inside and outside a superconductor?" No, this stream should be confined within a superconductor. Look at the flow of u in Fig. 101. The flow in this figure cannot be closed within the superconductor, it has no choice but to go outside it. Phase distributions such as those illustrated in Fig. 101 are therefore not possible.

Some of you might think, "The flow of Cooper pairs cannot go outside a superconductor, but isn't it possible for the flow of A to go outside?" Yes. But the problem in question is not the flow of A but that of $\frac{q}{m}A$. The vector potential A may exist outside the superconductor but the flow of $\frac{q}{m}A$ does not since there are no Cooper pairs outside it (q: electric charge of Cooper pairs). Therefore the flow of u exists only inside the superconductor.

If the shape of the phase distribution in a superconductor is that of a mountain like that in Fig. 101, it is evident that a flow must inevitably be produced, since the gradient of the phase (mountain slope) does not vanish. However, if the flow of u — or the water falling down the mountain — has neither sources nor sinks and is confined within the superconductor, we can have no idea how the water is produced and where the water goes. There can be no such streams.

We thus conclude that there is only one solution: the phase distribution should have neither mountains nor valleys; it should be flat. This means the flow of u has to vanish everywhere inside the superconductor. There is no solution other than a constant phase inside a superconductor. As long as a superconductor is shaped like a lump without any holes (*i.e.* is singly-connected), the flow of u always vanishes inside it.

To summarize, when a magnetic field is applied to a lump of superconductor there are two cases that can occur: (1) The applied magnetic field penetrates the superconductor, thus breaking down the superconducting state to become normal, or (2) the phase of the wave of Cooper pairs becomes constant, forming the Meissner state, about which we will consider in the next two sections of this chapter. (To tell the truth, it will be shown later that there is the third state, the vortex state.)

Circulating Current Around a Superconductor

At last we have arrived at the point where we can start the explanation of the Meissner effect. Although we considered a lot about what happened when a magnetic field was applied to a superconductor, we were led to a very simple conclusion: Even when a magnetic field is applied to a superconductor, $u = v + \frac{q}{m}A$ has to vanish everywhere inside it in order to maintain the superconducting state.

"Does the velocity v of Cooper pairs then also have to vanish?" Not necessarily. Even if v is not zero, u can be made zero if A satisfies the following equation:

$$v = -\frac{q}{m}A.\qquad(32)$$

This equation, however, cannot always hold, but only in special cases. For example, if Cooper pairs flow everywhere inside the superconductor ($v \neq 0$), a nonzero vector potential A should exist everywhere inside it, and the situation becomes similar to that when a magnetic field penetrates a superconductor. The superconducting state will be broken.

"What are special cases?" you might ask. One example is that Cooper pairs circulate ($v \neq 0$) in the surface layer of a superconductor without breaking down the superconducting state; this produces the Meissner effect.

Some of you might interrupt me to say, "You said that all the flows of u, v, and A were just like water streams falling down the mountain (see Fig. 101). What kind of the phase distributions can we imagine for Cooper pairs circulating around the surface layer of a superconductor?" You are to the point. There can be no such mountain slopes that a water stream continues to flow round and round. If we are forced to draw such a slope of the phase, it will become like a spiral staircase, as illustrated in Fig. 102(a).

This phase slope is quite strange: if we trace the phase of the wave starting from one point in the surface layer, going around a circle in the surface layer and returning to the original position, then the phase value becomes different from the original one.

"Can the wave of Cooper pairs exist in such strange situations?" The wave does exist! Look at the illustration in Fig. 102(b). The phase distribution caused by vector potential A also draws a spiral staircase but just in the opposite rotation direction. Therefore the summation of the two phase distributions results in a flat phase distribution, since $u = v + \frac{q}{m} A = 0$. If we consider only the phase distribution

Surface current

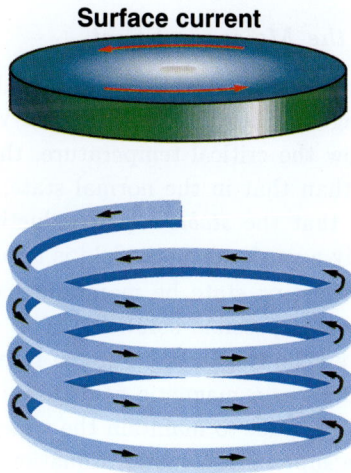

Phase distribution due to the flow of Cooper pairs (v)

(a)

Fig. 102. Phase distribution of the Cooper-pair wave in the Meissner state.
(a) Circulating supercurrent (qv) producing the phase distribution drawing a spiral staircase
(b) Circulating vector potential (A) producing just the opposite phase distribution due to v

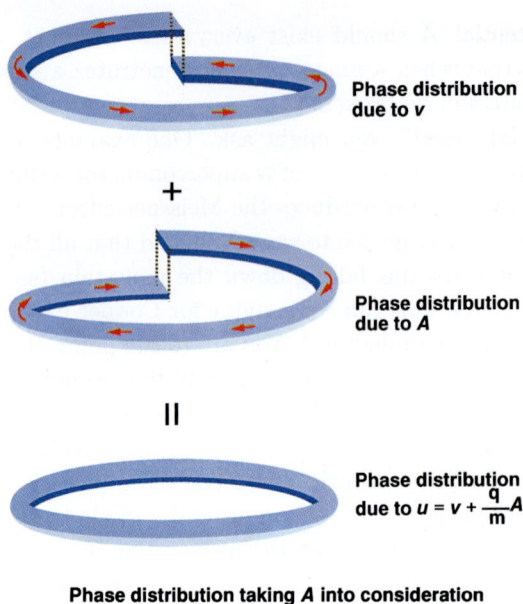

Phase distribution due to *v*

+

Phase distribution due to *A*

||

Phase distribution due to $u = v + \frac{q}{m}A$

Phase distribution taking *A* into consideration

(b)

Fig. 102. (*Continued*)

of either ***v*** or ***A*** we find strange things happening, but the phase distribution in total produces nothing strange.

Solving the Mystery of the Meissner Effect

We could finally get a clue in answering the question about what happens to a superconductor when a magnetic field is applied to it. When the temperature of a superconductor falls below the critical temperature, the superconducting state, which has a lower energy than that in the normal state, is produced. Therefore, it does not seem probable that the *stable* superconducting state is broken down immediately even when only a weak magnetic field is applied.

"How can the superconducting state be maintained when a magnetic field is applied?" We already knew a possible solution. The superconducting state is maintained by a supercurrent flowing in the surface layer of the superconductor. This circulating current produces a magnetic field that cancels the magnetic field inside the superconductor. In order to maintain the superconducting state also in the surface layer where the supercurrent flows, Equation (32), $v = -\frac{q}{m}A$, has to be satisfied. That means that a magnetic field has to exist at the surface layer. When we view this from the opposite angle, a magnetic field penetrates the surface layer of a superconductor inducing a supercurrent to maintain the superconducting state, which excludes the magnetic flux from inside the superconductor.

In this way, the Meissner effect has at last been explained. Some of you may not be fully convinced. "The Meissner effect can possibly be produced, but there is no guarantee that such an effect is actually produced." We need to use mathematical equations for such detailed explanations, but I will only sketch the outline of the derivation here.

Up to now we have been thinking only about the behavior of the wave of Cooper pairs, but from only this we cannot know the distribution of the magnetic field inside and outside a superconductor when a magnetic field is applied to it. We also have to solve Maxwell equations, which determine the behavior of electromagnetic fields. Usually we use Maxwell equations to derive electromagnetic fields, here vector potentials A, from electric currents qv. But this time both A and v are not given. "How can we get solutions in such a case?" We can get them by using both Maxwell equations and Equation (31), $v = -\frac{q}{m} A$, derived from the law of superconductivity.

The result, which can be obtained without any difficulty if you have experience solving Maxwell equations, is simple: the vector potential A penetrates only a thin layer at the surface of a superconductor, and Cooper pairs flow only in the layer in order to make u vanish. The supercurrent, the flow of Cooper pairs, in the surface layer produces a magnetic field to cancel the magnetic field inside the superconductor. This is the Meissner effect.

Here let us summarize the Meissner effect again. The properties of the wave of Cooper pairs are such that its phase is constant everywhere inside a lump of super-conductor even when a magnetic field is applied. Otherwise, the wave of Cooper pairs cannot exist and is destroyed. When we solve Maxwell equations under this condition of the constant phase of the wave, we find that a supercurrent flows only in the surface layer of the superconductor. This supercurrent produces an additional magnetic field such that the total magnetic field inside the superconductor vanishes. A vector potential penetrates the thin surface layer to cancel the phase shift caused by the supercurrent keeping the phase of the wave of Cooper pairs constant everywhere inside the superconductor. What an ingenious mechanism a superconductor has!

It is a matter of course that this supercurrent flows without any resistance. A superconductor is like a gigantic atom, and an atom cannot exist if there is any resistance that impedes electrons circulating around the nucleus. The supercurrent flows not because a voltage difference is applied, but because a magnetic field exists inside the superconductor. The vector potential A produces a gradient of the phase distribution, a gradient that makes the wave of Cooper pairs move to maintain the superconducting state.

Before leaving the Meissner effect, I would like to touch upon the gauge of vector potentials in superconductors. I have explained up to now that when a magnetic field is applied to a superconductor a supercurrent given by $-\frac{q}{m} A$ flows in the surface layer of the superconductor. The vector potential A can now be interpreted as a current and can be observed! In fact, the circulating supercurrent can be observed as the magnetic field. Here, A should have no freedom of gauges; gauge symmetry is broken in superconductivity.

It was in 1957 that the Meissner effect was elucidated in this way, by using the wave of Cooper pairs. Four years later a phenomenon called *magnetic flux quantization* was discovered experimentally, which the phase of the wave of Cooper pairs plays the main role in inducing.

Discovery of Flux Quantization

On July 15, 1961, four historical papers appeared in *Physical Review Letters*. The Americans B. S. Deaver and W. Fairbank and the Germans R. Doll and M. Näbauer had independently discovered *magnetic flux quantization.*

Both groups had fabricated hollow superconducting cylinders 10 μm in diameter and found that magnetic fluxes trapped in the holes of the hollow cylinders had discrete values (were quantized) (see Fig. 103). It is interesting to note that Deaver and Fairbank interpreted their results as the quantization of the magnetic flux in $h/2e$ units, whereas Doll and Näbauer stated "the magnetic flux unit measured was only 40% of the theoretical value of h/e, the reason being unknown."

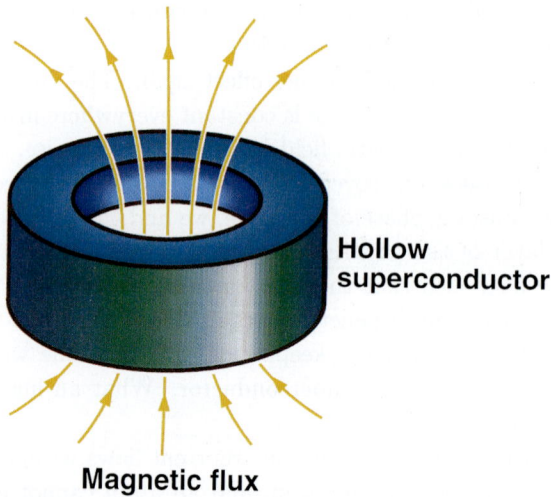

Hollow superconductor

Magnetic flux

Fig. 103. Magnetic flux trapped in a hollow superconducting cylinder.
The trapped magnetic flux was demonstrated to be an integral multiple of $h/2e$.

"Why is a magnetic flux trapped in a hollow superconducting cylinder quantized?" It was F. London who first predicted the possible quantization of a magnetic flux for the first time. He did not seem to have much confidence in the prediction, however, since it appeared only as a footnote in his famous 1948 book, *Superfluidity*. Long before the mystery of superconductivity was unraveled by using the wave of Cooper pairs, London explained superconductivity as follows: electrons in a superconductor must behave as a coherent wave. The phase of the electron wave in a hollow superconducting cylinder changes when the wave goes around the cylinder through the hole of which a magnetic flux passes. After one turn the phase shift is given by $\frac{2\pi q}{h}$ times the magnetic flux. However, the wave cannot exist unless the phase shift produced after one turn vanishes, or is equal to an integral multiple of 2π He therefore thought that magnetic flux surrounded by a superconductor must be quantized in h/e units. Since Cooper pairs had not yet been discovered, he had no choice but to think that *electrons* somehow formed the wave.

"How can we interpret this flux quantization in $h/2e$ units?" This question was answered by the two theoretical papers that appeared in the same journal.

Flux Quantization Requires New Physics?

C. N. Yang was visiting Stanford University when Deaver and Fairbank got their first experimental results, and he told me he was not convinced by the data Deaver and Fairbank showed him. But the new data he saw on his next visit convinced him that flux quantization had indeed been demonstrated. He was impressed with the insight of the experimentalists.

Yang and N. Byers at Stanford University considered the problem, "Does the magnetic flux quantization require new physics?" They concluded that the flux quantization could be explained without any new physics and that the quantization in $h/2e$ units provided evidence that superconductivity was produced by the pairing of electrons. The other theoretical paper was by L. Onsager, who had in fact predicted the flux quantization seven years earlier: he thought that in order for the electron wavefunction to have a definite phase value at a point in a hollow superconducting cylinder, the magnetic flux had to be quantized in h/e units.

The discovery of magnetic flux quantization made it vividly clear that superconductivity was a phenomenon in which the effects of quantum mechanics, thought to govern only the microscopic world of atoms and molecules, could also become evident in the macroscopic world.

Supercurrent Flowing in a Hollow Cylinder

Quantized magnetic fluxes trapped in hollow superconducting cylinders must have been produced by circulating supercurrents. Let us consider how supercurrents could be produced in those experiments.

The process of the experiments is as follows. When a magnetic field is applied to a hollow superconducting cylinder whose temperature is above T_c, the magnetic field passes through the superconductor without any disturbance [Fig. 104(a)]. When the cylinder is then cooled below T_c, the magnetic flux inside the superconductor is excluded by a circulating current in the surface layer (the Meissner effect). The hollow cylinder, however, has two surfaces, inner and outer. "Where does a supercurrent flow?" First of all, a supercurrent will flow in the outer surface layer of the hollow cylinder to exclude all the magnetic flux in the region inside the outer surface. If the superconductor is shaped like a lump without any hole, no other currents are induced. In the present case, however, there is also a surface at the hole.

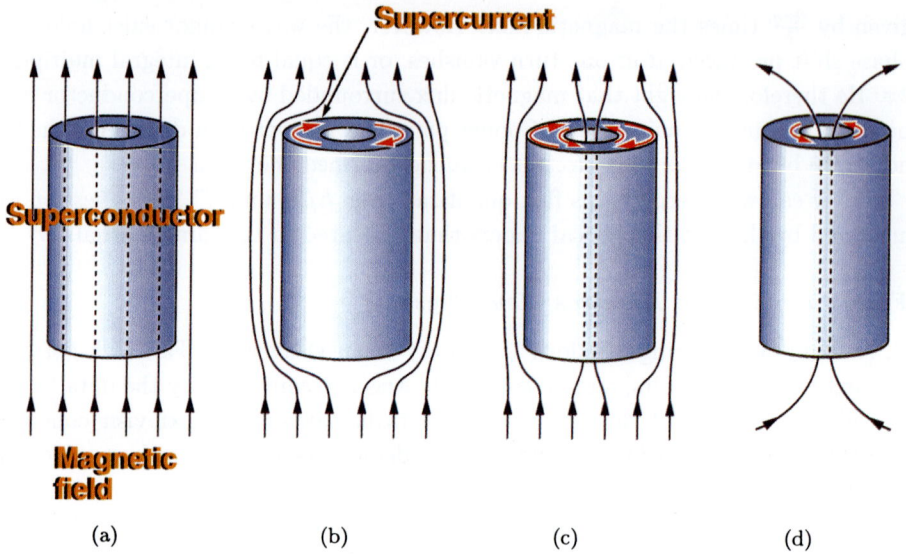

Fig. 104. Magnetic flux quantization
(a) Magnetic field passing through a hollow cylinder at $T > T_c$
(b) Supercurrent induced at the outer surface of the cylinder at T at $T < T_c$
(c) Supercurrents induced at both inner and outer surfaces
(d) Magnetic flux trapped in the hole of the cylinder when the magnetic field is removed

"Does a supercurrent also flow at the inner surface of the cylinder?" Some of you might think that no supercurrent should flow at the inner surface since the supercurrent at the outer surface has already excluded the magnetic flux inside it and there is thus no need for another supercurrent at the inner surface. This situation is illustrated in Fig. 104(b). Others might instead think that the Meissner effect should exclude a magnetic flux only inside a superconductor. That is, you might think that a magnetic flux inside the hole of the cylinder should not vanish. This situation is illustrated in Fig. 104(c): since the magnetic flux inside the hole has already vanished because of the supercurrent circulating in the *outer* surface layer, an additional supercurrent has to flow in the direction opposite to that of the outer supercurrent in the inner surface layer to produce a magnetic flux inside the hole.

"Which state actually occurs, (b) or (c) in Fig. 104?" Up to now we have considered the increase in energy only due to supercurrents, and consequently you may think that state (b) is more stable where a supercurrent flows only in the outer layer. When you think of the effect of magnetic fields, however, you will come to think that state (c) is more probable. This is because magnetic lines of force in the case (c) are almost straight and do not need to make a long detour around the cylinder, as they must in the case (b).

Although it seems that either case is possible, we will be able to determine a more stable state by calculating the energy of each state. However, judging from the experiment that a magnetic flux is trapped as shown in (d) after removing the applied magnetic field, the case (c) must be more stable. If the case (b) is assumed to occur, no magnetic flux should be trapped after removing the magnetic field.

The case (b) will occur when the magnetic flux passing through the hole at $T > T_c$ is smaller than $h/4e$, since it is the nearest quantized state.

The Wave of Cooper Pairs in a Hollow Cylinder

We have understood that when a magnetic flux is trapped in a hollow superconducting cylinder, a supercurrent flows in the inner surface layer as well as in the outer surface layer.

Next, let us consider the state of the wave of Cooper pairs not in the surface layer but well inside the superconductor. If no magnetic flux is trapped inside the hole, the phase of the wave of Cooper pairs is constant inside the superconductor. The problem is what happens when a magnetic flux is trapped inside the hole. In general, vector potential A circulates around the magnetic flux, as we studied in the case of the AB effect. A can penetrate even the superconductor [see Fig. 105(a)].

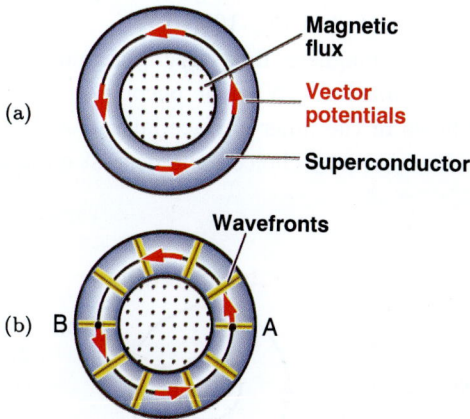

Fig. 105. Wavefronts of the Cooper-pair wave in a hollow cylinder trapping a magnetic flux.
    (a) Circulating vector potentials
    (b) Wavefronts
    A gradient in the phase distribution is produced by A. The wave can exist only when the phase after one turn is $2n\pi$.

"Is it possible to maintain the superconducting state when A exists inside the superconductor?" We verified that when a magnetic field B, and consequently A, penetrated a superconductor, the superconducting state would in general be destroyed. Please recall the reason for it. The reason comes from the fact that a phase shift of the wave of Cooper pairs after one turn along a closed loop is proportional to the magnetic flux enclosed by the loop (see Fig. 100). In the present case, however, vector potential A exists inside the superconductor but not magnetic field $\boldsymbol{B}$. The situation is quite different from that in Fig. 100 but is rather similar to the case of the AB effect (see Fig. 77).

Two kinds of closed loops can be drawn inside a hollow cylinder. One kind of loops does not enclose the magnetic flux and consequently produce no phase shift. The second kind of loops encloses the magnetic flux and produce a phase shift proportional to the flux enclosed. The superconducting state will be maintained only when a phase shift produced after one turn along the second kind of loops vanishes or is an integral multiple of 2π, that is, only when the flux trapped inside a hole of a superconductor is quantized in $h/2e$ units.

To understand this in more detail, let us consider the phase shift of the wave along a circular path [see Fig. 105(b)] starting from point A in the interior of the superconductor, tracing the circle in the counterclockwise direction and returning to point A via point B. Since vector potential $\boldsymbol{A}$ is circulating around the magnetic flux even inside the superconductor, the phase of the wave of Cooper pairs changes at a constant rate along the path. The phase shift after one turn must be zero or an integral multiple of 2π in order to maintain the superconducting state. That means the magnetic flux inside the hole of the cylinder has to be quantized in $h/2e$ units. If the flux is given by $nh/2e$ (n: integer), the phase difference becomes $2n\pi$. The case of $n = 10$ is illustrated in Fig. 105(b).

"What happens when a magnetic flux passing through the hole of a cylinder at $T > T_{\rm c}$ is different from $nh/2e$?" When it becomes superconductive at $T < T_{\rm c}$, such a supercurrent flows in the inner surface layer that the trapped flux may be quantized to $nh/2e$. Among many choices, the nearest quantized value of magnetic flux is chosen.

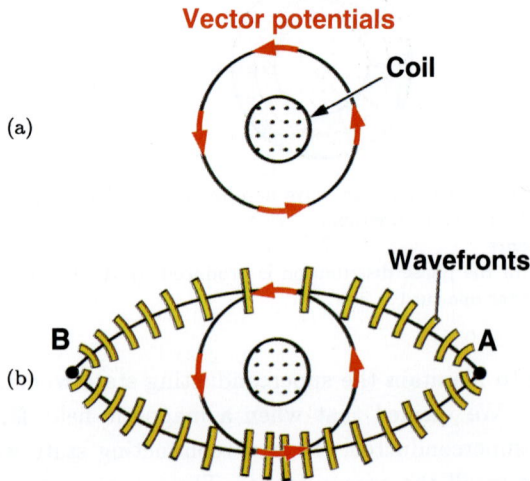

Fig. 106. Wavefronts of electron waves travelling from A to B on both sides of a coil.
(a) Vector potentials
(b) Wavefronts
The electron wave can exist even when the phase difference is not equal to $2n\pi$.

The flux quantization does not take place in the case of the AB effect where two electron beams enclose a magnetic flux to form an interference pattern [Fig. 106(b)], since electron waves do not form a continuous wave surrounding the magnetic flux. However, in the case of a superconductor the magnetic flux should be quantized so that the wave of Cooper pairs has to form a continuous wave.

Wavefronts of the wave of Cooper pairs are illustrated in Fig. 105(b). When no magnetic flux is trapped, the phase is constant everywhere in the superconductor. All of A, v, and u vanish. When a magnetic flux of $nh/2e$ ($n \neq 0$) is trapped, however, vector potential A circulates around the flux even inside the superconductor, thus producing a phase shift of the wave of Cooper pairs. Since the flow of A is concentric, the wavefronts are produced in the radial directions as in Fig. 105(b), which illustrates the case $n = 10$.

"Then how do the wavefronts move?" With time, the radial wavefronts rotate around the central axis of the cylinder. The direction of the wavefront movement is given by that of A.

Up to now, we have considered the wave of Cooper pairs only well inside the interior of the hollow superconducting cylinder where no supercurrent flows ($v = 0$). We would like now to think about the wave of Cooper pairs in the inner surface layer where a supercurrent is flowing ($v \neq 0$). "Can the wave of Cooper pairs in this surface layer also exist as a coherent wave?"

Supercurrent in Surface Layer

We already know that in the Meissner state the phase of the wave of Cooper pairs is a constant everywhere inside a lump of superconductor and the superconducting state is maintained everywhere inside the superconductor. Although Cooper pairs do move in the surface layer ($v \neq 0$) producing a phase gradient, the effect of v on the phase of the wave of Cooper pairs is canceled by vector potential A. However, the situation is different in a hollow cylinder trapping a magnetic flux: the phase of the wave is no longer constant but changes along a circular path around the magnetic flux.

We learned that this is valid only in the interior of the superconductor where no supercurrent flows, *i.e.* $v = 0$. "What about in the inner surface layer where a supercurrent is flowing ($v \neq 0$)?" Let us consider a circular path starting from A and returning to A via B as shown in Fig. 105(b). "What happens when the circle becomes smaller to pass through the surface layer where a supercurrent is flowing?" Since the magnetic flux trapped in the hole of the cylinder is produced by the supercurrent flowing in this surface layer, the magnetic flux enclosed by a circular path in this layer becomes smaller than $nh/2e$. That means the phase shift of the wave of Cooper pairs due to A is no longer $2n\pi$ but is smaller than that. If that was really the case, the wave of Cooper pairs would be destroyed. The wave survives, however, because of an additional phase shift produced by the supercurrent qv so that the total phase shift remains to be quantized as $2n\pi$.

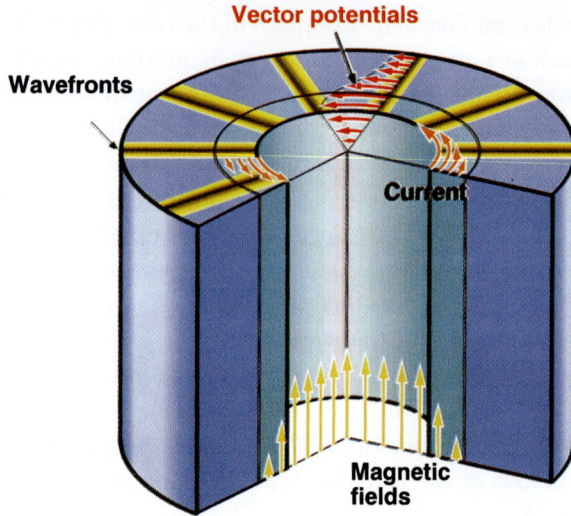

Fig. 107. Supercurrents and vector potentials in a hollow superconducting cylinder trapping a quantized magnetic flux.
Supercurrents flow only in the inner surface layer but circulating vector potentials exist even inside the superconductor.

Look at the illustration in Fig. 107. A supercurrent flows in the inner surface layer forming a current-carrying coil. The magnetic field is produced not only inside the hole but penetrates the surface layer. The total magnetic flux is quantized as $nh/2e$. The direction of vector potential A is tangential to concentric circles around the central axis. The value of A increases with the distance from the axis, and begins to decrease outside the inner surface. Wavefronts extend in the radial direction from the central axis, and wavefronts after one turn are returned to the original positions of the wavefronts. Even in the inner surface layer of a superconductor where a supercurrent flows, the superconducting state is maintained and wavefronts are also in the radial direction.

It has been clarified that the wave of Cooper pairs can exist as a single continuously connected wave everywhere inside a superconductor, whether a lump or a hollow cylinder, even when a magnetic field is applied. To maintain the superconducting state, supercurrents flow in the surface layers of superconductors producing the Meissner effect in one case and the magnetic flux quantization in others.

Observing the Flux Quantization Process

Deaver, Fairbank, Doll, and Näbauer demonstrated that the magnetic flux trapped in a hollow cylinder was always quantized in $h/2e$ units regardless of the magnitude of the applied magnetic field. This was interpreted as follows: when the wave of Cooper pairs inside the hollow cylinder notices that the wave once turned around it is not in phase with the original wave, a supercurrent is made to flow so that the two waves may be in phase or differ by $2n\pi$.

"Can we observe the behavior of a supercurrent when it begins to flow to quantize the magnetic flux?" This is the experiment that C. N. Yang proposed as a confirmation experiment of the AB effect which I mentioned in Chapter 10.

The experimental arrangement is a little bit different from the flux quantization experiment hitherto explained: first, the sample is not a simple superconducting cylinder as shown in Fig. 104, but is a hollow superconducting ring as shown in Fig. 90. Second, an external magnetic field is not applied to the sample but a magnetic field is applied only within the hollow of the superconducting ring in the form of a magnet placed inside the hollow (see Fig. 90). When the sample is cooled below T_c, a supercurrent begins to flow at the inner surface layer of the hollow so that the magnetic flux inside the hollow is quantized. Of the many possible quantized states, the state requiring the minimum supercurrent is selected. Since the magnetic field does not change with time, we cannot attribute the flux quantization to electromagnetic induction. The magnetic field B passes through only inside the hollow of the superconductor and does not touch the interior of the superconductor. We have no choice but to consider that the wave of Cooper pairs gets information about the magnetic flux by direct interaction with vector potentials A. This phenomenon can be said to be the Cooper-pair version of the AB effect.

It may be easier to consider if we take concrete examples. Let us assume the magnetic flux passing through the hollow is initially 8.8 $(h/2e)$. When the sample becomes superconductive, an additional supercurrent that produces a magnetic flux of 0.2 $(h/2e)$ flows in the inner surface layer of the hollow. The total flux becomes 9.0 $(h/2e)$.

"What happens when a magnetic flux inside the hollow of the sample at the normal state is 10.2 $(h/2e)$?" This time, an additional supercurrent flows in the opposite direction so that the total magnetic flux may be reduced by 0.2 $(h/2e)$, thus resulting in 10 $(h/2e)$. In this way, the magnetic flux jumps to the nearest quantum state. The moment of this transition can be captured by electron interferometry.

The Moment of Flux Quantization!

The sample here is a ring magnet made of permalloy completely covered with a superconducting niobium layer, the same sample used for testing the AB effect and shown in Fig. 90. An electron beam is incident onto such a ring and the phase difference between the beam passing through the ring hole and the beam passing outside the ring is measured in the form of an interferogram while the sample temperature decreases from above T_c to below T_c.

The experimental results are shown in Fig. 108. The interferogram at room temperature is shown in Fig. 108(a), where the interference fringe inside the hole is only a little bit above the outside fringe. The displacement corresponds to a phase difference of 0.3π. Since according to the AB effect, a 2π phase shift is produced between two electron beams enclosing a magnetic flux of h/e, the magnetic flux flowing inside the hollow of the ring must be 8.3 $(h/2e)$. From this photograph it

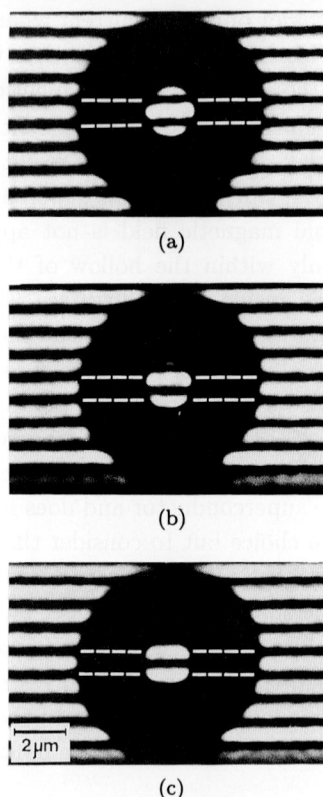

Fig. 108. Observation of flux quantization using a ring magnet covered with a superconductor.
    (a) Phase difference $= 0.3\pi (\text{mod} \cdot 2\pi)$ at 300 K
    (b) Phase difference $= 0.8\pi (\text{mod} \cdot 2\pi)$ at 15 K
    (c) Phase difference $= \pi (\text{mod} \cdot 2\pi)$ at 5 K
    The phase difference between the electron beam passing inside the ring hole and the beam passing outside the ring gradually increases with a decrease in the temperature T. When T decreases and crosses T_c, the phase difference suddenly jumps to π due to the flux quantization as shown in (c).

can only be known that the flux is $(n+0.3)(h/2e)$. The fact that $n = 8$ was inferred from the measured area of the cross section of the permalloy ring magnet.

The sample is first gradually cooled to 15 K $(-258°\text{C})$. The phase difference increases to 0.8π as shown in Fig. 108(b), meaning the flux has increased to $8.8(h/2e)$. "Why does the magnetic flux increase?" Since the temperature is still 15 K and above T_c (9.2 K), the covering niobium layer is not yet superconductive. The reason for this increase in magnetic flux is that the thermal fluctuations of spins inside the magnet decrease. Although spins are aligned in a domain of a magnet, the spin directions are always fluctuating because of thermal energy. Therefore, the magnetic flux consisting of spins becomes larger when the temperature is lower.

When the temperature decreases below 9.2 K, the fringe inside the hole suddenly jumps to the middle line between two outside fringes [see Fig. 108(c)]. "What

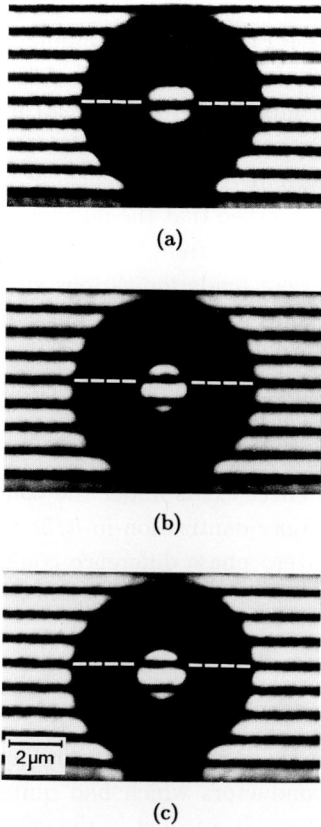

(a)

(b)

(c)

Fig. 109. Observation of flux quantization using a ring magnet having a different magnetic flux from that in Fig. 108.
(a) Phase difference $= -0.2\pi(\text{mod} \cdot 2\pi)$ at 300 K
(b) Phase difference $0.3\pi(\text{mod} \cdot 2\pi)$ at 15 K
(c) Phase difference $= 0(\text{mod} \cdot 2\pi)$ at 5 K
In this case, the phase difference vanishes due to the flux quantization.

happened?" We captured the moment the magnetic flux was quantized. The superconducting layer cannot enclose a fractional flux of 8.8 $(h/2e)$ within it without breaking down the superconducting state. Cooper pairs would like to maintain the more stable superconducting state, and for that purpose, a supercurrent flows so that the total magnetic flux increases to 9.0 $(h/2e)$. We observed the mystery of the wave of Cooper pairs! The wave of Cooper pairs knew the value of the magnetic flux through vector potentials, since the magnetic field did not touch the superconductor.

Now let us see another sample (Fig. 109). The fringe inside the hole is located below the outside fringe by a phase difference of 0.2π, as can be seen from Fig. 109(a). The magnetic flux of 9.8 $(h/2e)$ is enclosed in this ring magnet. When the sample temperature decreases to 15 K, the flux increases to 10.3 $(h/2e)$, as shown in

Fig. 109(b). When the temperature falls below T_c the magnetic flux jumps to a quantized value, 10 $(h/2e)$ [see Fig. 109(c)] by inducing a supercurrent to flow in the direction opposite from that in the previous case.

"Why does the magnetic flux jump to 10 $(h/2e)$ instead of 11 $(h/2e)$?" The reason is simple: the flux quantization to 10 $(h/2e)$ can be attained with a smaller amount of additional supercurrent.

In these experiments, we observed that the magnetic flux jumped to the nearest quantized state by inducing a supercurrent.

Although measurements were made for various sizes of samples, as shown in Fig. 88, there were only two kinds of results: the phase difference between two electron beams was either π [Fig. 108(c)] or 0 [Fig. 109(c)]. Since this flux quantization occurs only when a magnetic flux is completely surrounded by a superconductor, the quantization of the phase difference assures that no magnetic fields leak outside the sample due to the Meissner effect. Therefore, only a single photograph in Fig. 91 tells us that the AB effect exists under the condition of vanishing leakage magnetic flux. That is, the flux quantization in $h/2e$ units indicates that there is no leakage flux, and the nonzero phase difference confirms the physical reality of vector potentials.

Abrikosov's Prediction

Superconductors Withstanding Strong Magnetic Fields

In 1957, A. A. Abrikosov in the Soviet Union, now in the U.S.A., predicted the existence of Type II superconductors which had quite different properties from those in Type I superconductors. It was just in the year when the BCS theory was presented, four years before the discovery of flux quantization. For such reasons or others, Abrikosov's prediction was not recognized soon. In 1961 superconductors withstanding strong magnetic fields were discovered, and to solve the riddle of why superconductors full of impurities withstand strong magnetic fields, the once-shelved Abrikosov's theory was taken out.

When a weak magnetic field is applied to a Type II superconductor, the magnetic flux is expelled as a result of the Meissner effect. When the magnetic field exceeds a critical value at some surfaces of the superconductor, the magnetic flux penetrates the superconductor in the form of thin filaments (*vortices*).

In the beginning of this chapter, a vortex was introduced without touching on the reason, "Why is the superconducting state broken down in the form of filaments?" or "Why is the magnetic flux of a vortex equal to $h/2e$?" Now that we know the existence of the continuously connected wave of Cooper pairs inside a superconductor, it is not difficult to understand why quantized vortices are produced.

We have already learned that when a magnetic field is applied to a superconductor with a hole, a magnetic flux having a flux $nh/2e$ can be trapped inside the hole. Now we need to understand, "Why are vortices produced in a superconductor having no holes?" and "Why is only a magnetic flux of $h/2e$ selected and not its multiples?"

Let us consider these problems from the energy point of view. In the first place, the fact that superconducting materials become superconductive below T_c means that the superconducting state is lower in energy than the normal state. The superconducting state is preserved when nothing particular happens. Even when a weak magnetic field is applied, the superconducting state is not broken down soon and is preserved, kept as the Meissner state. Since magnetic lines of force have to go around the superconductor [see Fig. 92(a)], the magnetic energy increases. When the applied magnetic field increases, a larger supercurrent needs to flow. Therefore, it is evident that there are limitations on this state.

"What happens when the "pressure" of the magnetic lines becomes intolerable?" Let us consider the loss and gain of the energy. If magnetic lines are assumed to penetrate some regions of the superconductor as shown in Fig. 92(b), the energy increases because the regions change from superconductive to normal while the magnetic energy decreases because magnetic lines become almost straight. Therefore, the penetration of magnetic lines may occur especially in the case of a thin film: when a magnetic field is applied perpendicular to the surface of the film, magnetic lines have to make a long detour [see Fig. 92(a)].

Here we have to take into account another effect: the effect of a boundary between normal and superconducting regions. When magnetic lines partially penetrate a superconductor, the boundaries are necessarily produced. Such a boundary is similar to the surface layer of a superconductor where a shielding current flows because of the Meissner effect. Although this layer is thin, it must have a great effect on tiny vortices.

"Is the boundary energy positive or negative?" The fact is that there are both cases. In a Type I superconductor, such as lead, the boundary energy is positive. In a Type II superconductor, such as niobium and high-T_c superconductors, the energy is negative. Consequently, in a Type II superconductor, the boundary region tends to be as large as possible. If magnetic fluxes are assumed to pass through a Type II superconductor, they would tend to be separated into fluxes having the minimum unit $h/2e$, thus forming vortices.

In contrast, magnetic fluxes pass through a Type I superconductor in such a way that the boundary region is as small as possible. That means magnetic fluxes tend to get together. In this way, the superconducting state is totally broken down in a whole sample. If the magnetic field is not strong enough to break down the superconducting state in the whole sample, magnetic fluxes partially penetrate the superconductor, producing what is called the *intermediate state*.

For example, when a magnetic field perpendicular to a film is applied, magnetic lines have to make a detour all around the film as shown in Fig. 92(a). The magnetic field at the film edge becomes so strong that the superconducting state breaks down there and magnetic fluxes penetrate there. When the magnetic field is not so strong as to bring the whole film region into the normal state, the film is divided into normal and superconducting domains. Such an intermediate state is captured in Fig. 97(b).

"How are vortices arranged in a Type II superconductor when a strong magnetic field is applied to it?" According to Abrikosov's theory, vortices are closely packed to form a lattice, since they repel each other.

Practical Superconductors

It took some time before the Abrikosov's theory of Type II superconductors was accepted. Once the mechanism of newly discovered superconductors, which could withstand strong magnetic fields, was clarified by this theory, similar practical superconducting materials were developed one after another. The powerful magnets Onnes dreamt of have become realities.

It was not only the production of vortices that made it possible to develop powerful magnets. The superconducting state is also destroyed when a large current is applied. "What is the reason for this breakdown?" Recall the illustration in Fig. 94. When a current is applied, the vortices are acted upon by the Lorentz force and begin to move unless they are pinned down. Moving vortices produce heat that breaks down the superconducting state. If the superconducting state is to be maintained, the vortices must be pinned down, or "fixed."

"Is there any way to pin vortices?" The secret for the vortex pinning was hidden in the materials of powerful magnets: the materials were not uniform but had tiny impurities within them. Since the impurity regions did not become completely superconductive, it was not easy for vortices once trapped there to escape from them. Tiny impurities acted as pinning centers.

It has been clarified in this way that vortices are not only related to the fundamentals of superconductivity but also hold a key to the practical application of superconductors. People began to think in due course that they wanted to observe the vortices Abrikosov's theory predicted.

Vortices Form Triangular Lattices!

The vortex lattice predicted by Abrikosov was directly visualized in 1967, when D. Essmann and H. Träuble in Germany made images of vortices by using the Bitter technique (Fig. 95). They sprinkled iron particles over a superconductor surface and used electron microscopy to observe the particles gathered at the vortices. A photograph taken by Essmann and Träuble is shown in Fig. 110. You can see that vortices form a triangular lattice.

Several attempts were made to directly observe vortices by transmission electron microscopy, but the deflection angle due to a vortex is less than 1/1000 degree and was too small to be detected with conventional electron microscopes. It was not until 1989 that *magnetic lines of force leaking outside from vortices* in a superconductor could be observed (see Fig. 97) by electron holography using a field-emission electron microscope we developed.

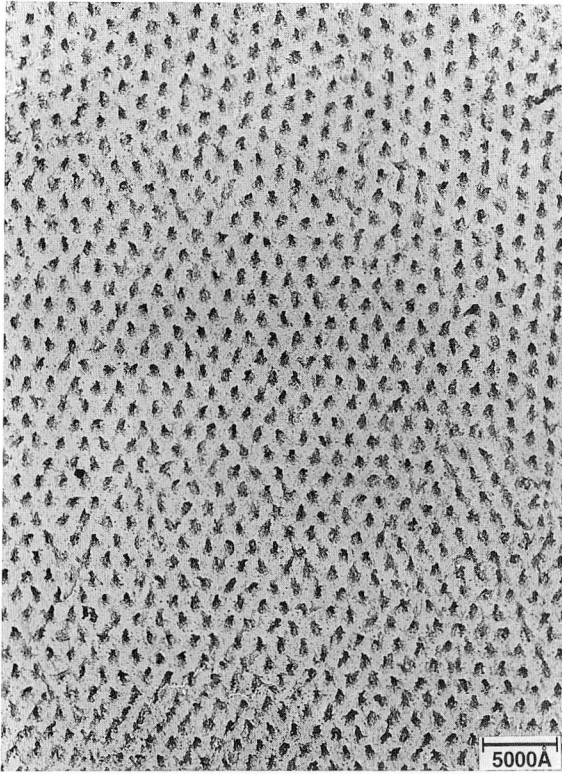

Fig. 110. A Bitter pattern of vortices penetrating a superconducting niobium (by courtesy of U. Essmann).

It is true that individual vortices could be observed in the form of magnetic lines of force. With this method, however, neither a two-dimensional array of vortices nor defects inside a superconductor could be observed. If vortices and defects were observed simultaneously, the flux pinning mechanism determining the critical current, or the interaction of vortices with pinning centers will become microscopically observable before our eyes, which has never been possible before.

Develop Techniques to Observe Vortex Dynamics
350 kV Holography Electron Microscope

In 1984, Hitachi, Ltd. celebrated its 75th anniversary. As part of the celebration, a new laboratory called *Advanced Research Laboratory* was planned to be established for long-range research. H. Watanabe, who initiated electron-holography research at Hitachi and was then a vice-president of Hitachi, led the efforts to create this laboratory.

We got the budget to develop a 350 kV holography electron microscope (Fig. 111). The microscope column was developed at the Naka works in Ibaraki,

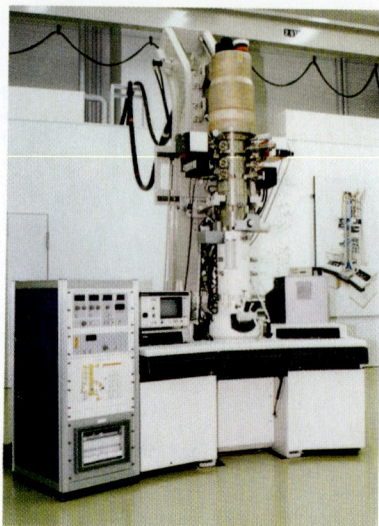

Fig. 111. A 350-kV holography electron microscope.

but the vital part of a field-emission electron gun had to be developed by our group members J. Endo and T. Kawasaki. Our research team at that time comprised six researchers, we had no choice but to devote our efforts to such a risky development project. I believe that if one wants to do new experiments, one has to develop the apparatus for them. One can even arrive at a target which has been thought to be a dream if one stands on the heights of technology. Therefore, I have always devoted half of our team to developing technologies and the other half to research.

When a new building was constructed at Hatoyama in Saitama Prefecture in 1989 (see Fig. 112), the microscope was at last completed and set up in an isolated building behind the main building. For three years Kawasaki used this microscope for research on high-resolution electron holography. At the end of March in 1992, when he left Japan to study at Cambridge University, we started research on vortex observation.

How to Observe Vortices

We racked our brains for ways to observe vortices directly and dynamically. After much thinking, we concluded that vortices could be observed in the experimental arrangement shown in Fig. 113. A superconducting niobium thin film is tilted and a magnetic field is applied horizontally. When the magnetic field is weak, magnetic lines of force go around the film because of the Meissner effect. When the field becomes stronger, magnetic lines penetrate the film in the form of vortices. An electron beam is incident from above and vortices are observed in an interference micrograph obtained through the holography process.

Fig. 112. Hitachi Advanced Research Laboratory, Hatoyama, Saitama.

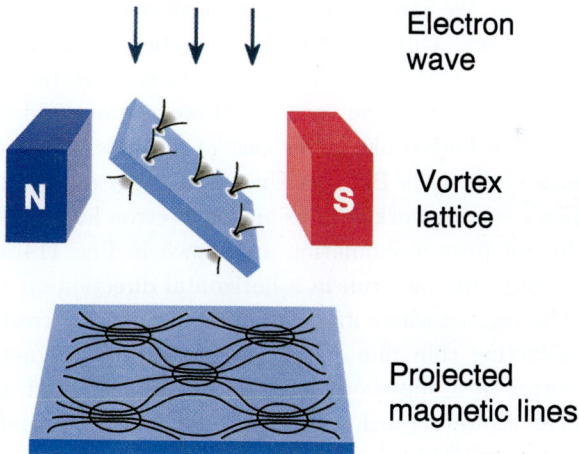

Electron wave

Vortex lattice

Projected magnetic lines

Fig. 113. Electron holography for observing magnetic lines of force of vortices in a superconducting thin film.

"Why does the thin film have to be tilted?" Samples in an electron microscope are usually placed horizontally and not tilted. In the case of the vortex observation, however, an electron beam is not deflected when a film is placed horizontally. The reason for it is that magnetic fields of vortices penetrate the film perpendicular to the film plane, and that the directions of an electron beam and the vortex magnetic fields become parallel.

"How can magnetic lines of force be seen in this arrangement?" Magnetic lines of force projected in the direction of an electron beam can be observed by

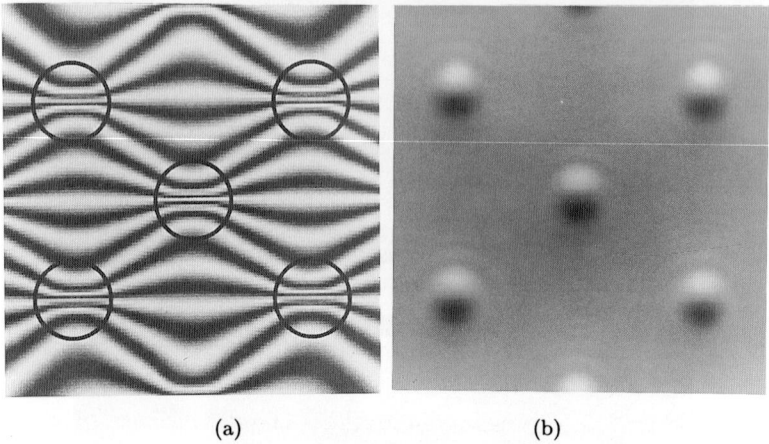

(a)   (b)

Fig. 114. Simulated images of vortices.
(a) Interference micrograph
(b) Lorentz micrograph

interference microscopy. We can easily imagine projected magnetic lines of force in the case of a single vortex. Outside magnetic lines tied up in bundles penetrate a superconducting film as a vortex from the film surface. Then they go outside from the other surface to fan out into the vacuum. However, we could not imagine how projected magnetic lines looked like in the case of multiple vortices.

I consulted with G. Pozzi in Bologna University, who stayed at our laboratory for two weeks twice a year to work with us on the electron holography experiments, and the results he got from a simulation are shown in Fig. 114(a). You can see in the figure that magnetic lines run in a horizontal direction on the average, but are modulated. The regions where magnetic lines are dense correspond to vortices inside a superconducting thin film. It appears that the magnetic lines leaking outside from a vortex in a superconductor do not fan out, but are absorbed by neighboring vortices. However, such a strange thing does not actually happen. The strange behavior of magnetic lines of force in the interference micrograph is due to the effect of the projection.

We started experiments in which we expected to observe magnetic lines of force like those shown in Fig. 114(a), but the first picture of vortices we got was the moving, pinball-like image shown in Fig. 114(b).

Vortices Moved!

In June 1992, only three months after starting the experiment, K. Harada, H. Kasai, J. Bonevich, and T. Matsuda noticed an extremely dim figure of something like pinballs moving in line when the sample temperature changed. At first, other people could hardly recognize the figure. But when the experiments were repeated, the image became clearer and clearer.

A few days after we were convinced that these pinballs must be moving vortices, the general manager of our laboratory, S. Asai, called a meeting of everyone concerned: the deputy manager A. Suzuki who made all the efforts to establish our laboratory and our machine; J. Endo who developed the electron microscope; the four who were lucky enough to have a chance to see moving vortices for the first time; S. Kubota, S. Matsunami, and M. Moriya who developed the low-temperature specimen stage; and people who prepared the samples. All these people worked together in accord; if not, we could not have arrived at this goal so soon. The videotape prepared for the meeting showed the results obtained just the day before. Everybody, holding his breath, gazed at the scene humankind could see for the first time.

I happened to recall in my mind two people who did especially much to establish this unique laboratory, like none anywhere else in the world, and who allowed us to develop our risky apparatus when no one could predict the kind of results we would obtain with it. One of these people was Watanabe, the former vice president of Hitachi and at that time the president of Hitachi Maxell Limited, who was supposedly rumored to have served concurrently as the general manager of our laboratory when it was established. Without his continuous hidden support up to that time, this day would not have come. The other was the former general manager E. Maruyama who completely believed in me, let me do anything I wanted to do, and helped me develop our electron microscope. I visualized in my heart the scene that they were with us, delighted to see the video. Everybody felt that something superb was being done. The development after this was extremely rapid. Vortices have really become pinballs.

Why Pinballs?

You may not understand why vortices, which should have been seen as magnetic lines of force, looked like pinballs. The picture in Fig. 114(a) is of anything but pinballs. The fact is as follows. Since we could find no figures of vortices in the experiment, following the proposal of Matsuda, we tried to use *Lorentz microscopy* to reveal the vortices: Lorentz microscopy is simply an out-of-focus imaging method and has been used for observing domain structures in magnetic thin films. The principle behind it is not a difficult one (see Fig. 115). Electrons transmitted through a vortex are slightly deflected by its magnetic field and displaced in a plane far behind the film. No electrons arrive at the region just below the vortex, while the intensity of electrons doubles where the deflected electrons arrive. Thus the vortex can be seen as a pair of bright and dark contrast features. The direction of the vortex magnetic field is provided by the line dividing the bright and dark regions. This image was also simulated by Pozzi, and is shown in Fig. 114(b).

An actual Lorentz micrograph of vortices is shown in Fig. 116. The sample is a niobium thin film at 5 K, or $-268°$C, under a magnetic field of 100 gauss. White pinballs in the picture are vortices. When you observe them carefully, you can see that each vortex image consists of black and white regions as expected.

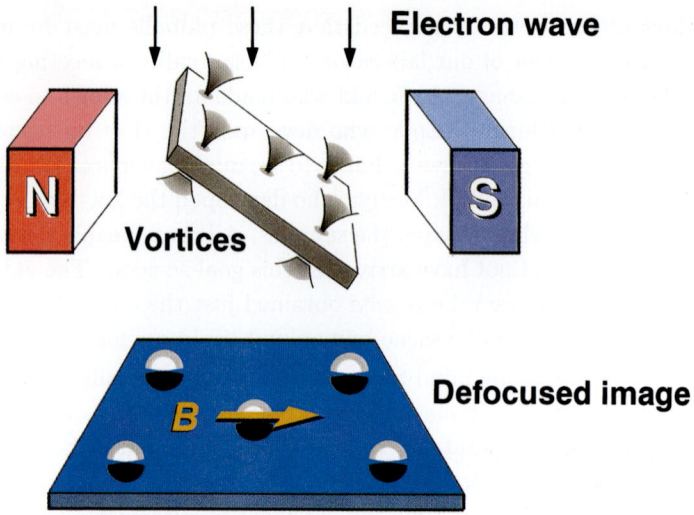

Fig. 115. Lorentz microscopy for observing vortices in a superconducting thin film.

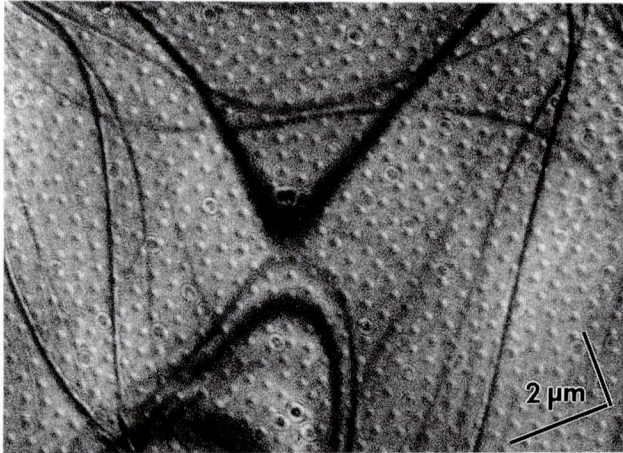

Fig. 116. A Lorentz micrograph of a vortex lattice in a superconducting niobium film.
Each spot with black and white contrast pairs corresponds to a vortex. Black fringes
indicate bend contours where an incident electron beam is Bragg-reflected.

In this way, we could first observe vortices not as an interference micrograph
but as a Lorentz micrograph. In due course, we were able to observe vortices
also by interference electron microscopy (see Fig. 117). Each of these two
methods has its advantages: with interference microscopy, magnetic lines of force
can be quantitatively observed in a high resolution image of a sample. Lorentz
microscopy, where vortices are observed as spots, is suitable for dynamic
observation.

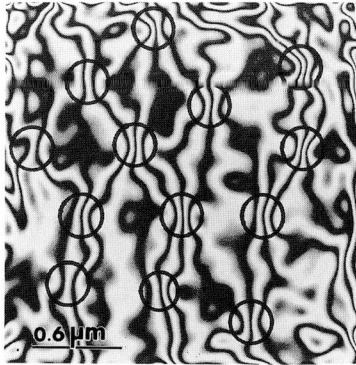

Fig. 117. An interference micrograph of a vortex lattice in a superconducting niobium film (phase
amplification: ×16).
Encircled regions in the photograph correspond to vortices.

Moving Vortices Look Like Living Creatures

Vortices observed by Lorentz microscopy looked like living creatures. They moved
in various manners when the magnetic field applied to a sample or the sample
temperature changed. The vortex movement was also interesting when there were
defects or impurities in the sample. Let me introduce you some examples. The first
shows what happened when the magnetic field applied to a superconducting niobium
film was suddenly removed. Most of the vortices seen in Fig. 116 disappeared, but
10% remained trapped at weak pinning centers. At first they looked still, but when
you looked at them carefully, they moved in an interesting way: vortices moved

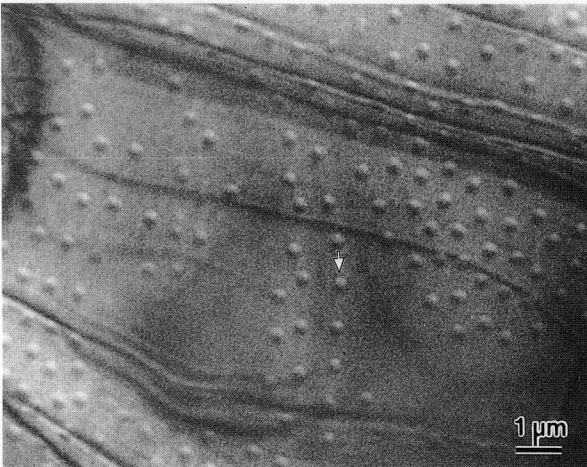

Fig. 118. A Lorentz micrograph of a superconducting niobium film at 5 K after a magnetic field
of 100 gauss is removed.
When the magnetic field is switched off, 90% of vortices disappear. The remaining
vortices look still, but gradually go outside the film by hopping in the downward direction
in the picture.

downwards in the picture towards the edge of the film as in Fig. 118. Vortices hopped among pinning centers as if they were jumping over stepping-stones. Vortices which happened to be released from pinning centers by thermal energy received a force toward the film edge. This force was due to the thickness gradient and also to the repulsion of magnetic lines of force of vortices inside the film. The released vortices hopped from one pinning center to another to go outside the film.

The next example shows the vortex behavior when the applied magnetic field was suddenly reversed. Original vortices wanted to go outside the film from the edge when the magnetic field was reversed, but doing so took some time. New vortices with the opposite orientation, however, could not wait and began to enter the film from the edge. The two kinds of vortices began to collide head on at some places. The photograph in Fig. 119 is the scene of a collision. It can be seen from the photograph that the two kinds of vortices are opposite-oriented, since

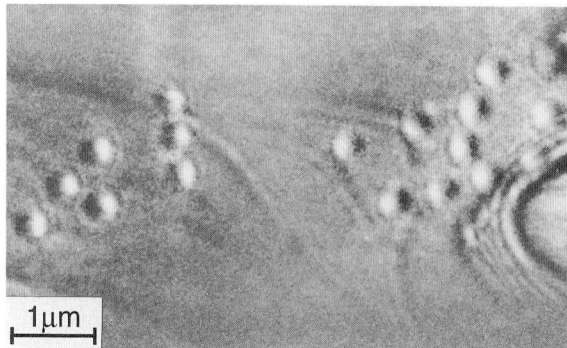

(a)

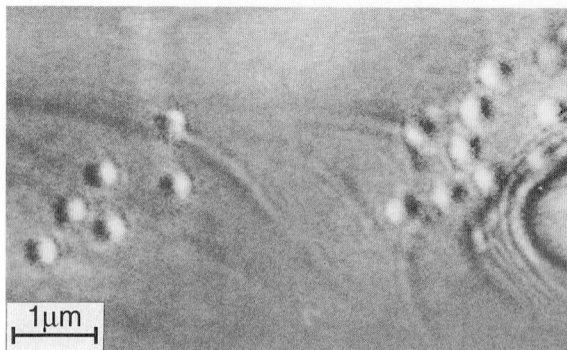

(b)

Fig. 119. Annihilation of antiparallel pair of vortices.
    (a) Before the annihilation
    (b) After the annihilation
    The directions of vortices on the right hand and left hand sides in the picture are opposite.
    When a vortex and an antivortex approach each other, they suddenly disappear.

the black side of the vortex image is opposite in both sides of the picture. Vortices in a superconductor are elementary particles in the sense that they cannot be divided any further. Therefore, we can look at the annihilation of particles and antiparticles. When two opposite-oriented vortices approached by hopping, they suddenly collided to disappear. The figures in Fig. 119(a) and 119(b) show the scenes just before and after the collision.

The last example shows the vortex behavior when they move in a superconducting thin film with surface steps (Fig. 120). When the value of the magnetic field applied to the film suddenly changed, vortices began to flow from the lower left to the upper right in the photograph. When vortices arrived at the surface step running in a vertical direction, they could not move across it because of the pinning effect and changed their direction to flow upward along the step. When they arrived at the intersection of steps, they again changed their direction, this time to the right. Since there were two horizontal steps, vortices sometimes moved along one step and sometimes along the other

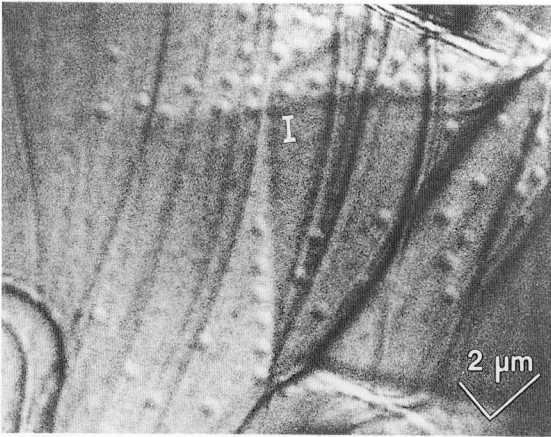

Fig. 120. Vortex movement in a superconducting niobium film with surface steps.
Vortices move from lower left to upper right when the applied magnetic field changes. When vortices reach the surface step running in the vertical direction of the picture, they cannot cross the step, and change their directions to move along it in the upward direction. When the vortices reach the intersection of steps designated as "I" in the picture, they turn right to move along two different steps.

We found that vortices moved in a variety of manners in the microscopic world inside superconductors. We are now able to actually observe what we previously only imagined. What I have introduced up to now is the movement of vortices in a superconducting *niobium* film. We are now attempting to obtain these motion pictures of vortices in high-T_c superconductors. Although the dynamic behaviors of vortices are thought to be the biggest reason that high-T_c superconductors cannot be put to practical use, the vortex behavior is still wrapped in mystery. Some people say that vortices fluctuate eternally like molecules in a liquid, forming a liquid phase under some conditions, and also that a vortex line in a superconductor is not straight but zigzag when a magnetic field is applied obliquely to the layers of high-T_c superconductors. We are now trying to make a new electron microscope, hoping that we can use it to show you such strange behaviors.

Epilogue

Thirty years have already passed since I started research on electron holography in 1967, the second year after I joined the Hitachi Central Research Laboratory. I have not devoted all the time to this research during this period, since the laboratory I belonged to was an industrial one. I have also worked on the developments of field-emission scanning electron microscopes and electron-beam lithography machines. When these developments finished, however, I had the opportunity to resume research on electron holography. Especially during the last ten years since the establishment of the Hitachi Advanced Research Laboratory, I have been able to research as much as I wish.

People often asked me how I could manage to continue to make fundamental research in an industry. I wonder, too. Although I can say with confidence that I have had a strong will to carry through something interesting by any means and have also made much effort second to none for that purpose, this may not have been enough. I think the reason I could do it lies in the fact that there have always been people close to me who warmly supported me: some people helped me directly and others stood by me without informing me of it.

Two years ago I met Dr. S. F. Ling of World Scientific at the Conference celebrating the Franklin Institute Medalists, including C. N. Yang. He was so eager to persuade me to write a book that I felt inclined to do so hoping that what I have done in my research may stimulate young people to feel interested in physics. When I read the completed manuscript, I had no confidence that my intention at the beginning was attained.

Since I could not find any time, I began to write the manuscript when I was on board international flights or when I could not sleep at hotels in foreign countries due to jet lag. Therefore, it took a long time. In addition, I often changed my mind. As a result, I caused my secretaries much trouble. I am grateful to Mses. M. Hotei, R. Fukuhara, and J. Kato for typing the manuscript and Mses. S. Kasai, T. Ishii and K. Matsuyama for preparing the figures.

I thank Prof. H. Ezawa of Gakushuin University and Dr. Y. A. Ono of Hitachi Advanced Research Laboratory for reading and criticizing the manuscript.

Index

Abrikosov A. A. (1928–), 144, 146
AB effect, 83, 85, 87–90, 97, 99, 102, 104–109, 113, 137, 138, 141, 144
aberrations, 13, 19, 20, 25, 26
Aharonov Y. (1932–) 83–86, 89, 90

Bardeen J. (1908–1991), 123
Bitter pattern, 76, 77, 118, 147
Boersch H. (1909–1986), 20
Bohm D. (1917–1992), 83–86, 89, 90
Bohr, Niels H. D. (1885–1962), 15
Bose particle(s), 59, 123, 125
brightness, 37, 38, 44
bubble domains, 3
Busch H. (1884–1973), 25

cathode ray, 14, 24
Chambers R. G. (1924–), 85
complex number(s), 53, 54, 56
conjugate image(s), 31–34, 64, 66, 68
contour map, 62, 77, 80
Cooper L. N. (1930–), 123
Cooper pair(s), 89, 123–131, 133–135, 137–141, 143, 144
correlation, 56–58, 60
Crewe A. V. (1927–), 36, 41
critical current, 115
critical magnetic field, 115
cross-tie wall, 77
cylindrical waves, 17

Davisson C. J. (1881–1958), 16, 17
de Broglie L. V. (1892–1987), 15, 16, 25

diamagnetism, 122, 123
Dirac P. A. M. (1902–1984), 16, 59

Einstein A. (1879–1955), 101, 102, 104
electromagnetic induction, 89, 92, 94, 117, 121, 122, 126
electron biprism, 21, 23, 35, 46, 47, 55, 62, 66, 84, 86, 104
electron holography, 24–35
electron microscope(s), 21, 25, 30, 35, 112
electrotonic state, 92–94
ether, 70
exchange force, 75, 80

Fairbank W. F. (1917–1989), 134, 135
Faraday M. (1791–1867), 1, 70, 92–94, 101
Fermi particles, 59, 123, 124
Feynman R. P. (1918–1988), 46, 85
fiber bundle theory, 104, 105
field-emission electron, 38, 42, 44, 59
flux line, 117
flux quantization, 109, 112, 113, 121, 127, 134, 135, 139–141, 144
fluxoid, 117
Fraunhofer holography, 33–35
Fresnel fringes, 20, 21, 36
Friday Evening Discourse, v, 1, 6

Gabor D. (1900–1979), 13, 24–26, 29–31, 35, 36

159